M & E HANDBOOKS

M & E Handbo
amination syllab
Handbook covers
in the series form ι
school and home s
  Handbooks con   ...рреа of
unnecessary paddiι.͜, ..ʌ..ιιg each title a comprehensive
self-tuition course. They are amplified with numerous
self-testing questions in the form of Progress Tests at the
end of each chapter, each text-referenced for easy
checking. Every Handbook closes with an appendix
which advises on examination technique. For all these
reasons, Handbooks are ideal for pre-examination
revision.
  The handy pocket-book size and competitive price
make Handbooks the perfect choice for anyone who
wants to grasp the essentials of a subject quickly and
easily.

THE M & E HANDBOOK SERIES

Human Geography

H. Robinson,

B.A., Ph.D., M.Ed.

*Head of Department of Geography,
and Environmental Studies,
and Dean of the Faculty of Arts,
The Polytechnic, Huddersfield*

THIRD EDITION

MACDONALD AND EVANS

Macdonald and Evans Ltd.
Estover, Plymouth PL6 7PZ

First published 1969
Reprinted 1971
Reprinted 1972
Reprinted 1973
Reprinted 1974
Second edition 1976
Third edition 1978

© Macdonald and Evans Ltd. 1978

7121 0813 0

Printed in Great Britain by
Hazell Watson & Viney Ltd,
Aylesbury, Bucks

Preface to First Edition

Human geography is one of the two main branches of modern geography, the other being physical. The term "human" obviously implies a relationship with man, and in human geography the emphasis may be said to be focused upon man—man in relation to his environment. Although the emphasis is upon man and his activities, this does not mean that the physical aspects of geography are ignored: they cannot be, for location, terrain, climate, vegetation and animal life all influence, for good or ill and in greater or lesser degree, man himself and the use he makes of his earthly domain.

Human geography is of importance because of its relevance to human activities and ideas and to modern life. As the world becomes increasingly smaller as a result of speedier transport and instantaneous communications, it is essential for us all to know something of other peoples and their problems. The present-day problems of exploding populations, food supplies, urban growth, world health and literacy as well as political antagonisms all need study and understanding. This HANDBOOK will, it is hoped, provide the first simple steps to this study and understanding.

Human geography itself, as will be seen in the first chapter, has several aspects and a very large book could easily be written upon each aspect. Here, in the small compass of this little book, it is obviously impossible to cover every aspect of the field of human geography and the author has had to be highly selective, touching upon one or two important topics. The needs of students, particularly those studying for examinations which include some study of human geography, have been kept more especially in mind and the topics dealt with are those commonly found in examination syllabuses.

This HANDBOOK gives some of the basic essentials of human geography but the student should familiarise himself with the specific coverage of his particular examination syllabus.

The HANDBOOK could form the basis of a course in human geography in Liberal Studies which are now common in Colleges of Further Education. Indeed, in the writer's own college courses in social geography have been undertaken for several years.

For a fuller treatment of many of the topics dealt with here the student is recommended to read the following books:

An Introduction to Human Geography, D. C. Money, University Tutorial Press.

An Introduction to Human Geography, J. H. G. Lebon, Hutchinson.

Human Geography, E. Jones, Chatto & Windus.

Frontiers of Geography, O. Hull, Macmillan.

A selection of questions taken from recent examination papers of the undermentioned bodies is included in Appendix II, and I am grateful to them for their permission to use these questions:

Institute of Bankers
Associated Examining Board
University of London
University of Oxford
Oxford & Cambridge Joint Board
Northern Counties Technical Examinations Council
Welsh Joint Education Committee.

A guide to the answering of examination questions is given in Appendix I. The student is advised to read this carefully and to heed what is said. Many years of experience as an examiner and chief examiner have shown quite clearly where students frequently fall into error and, all too often, do not do justice to themselves.

August 1969   H. R.

Preface to Third Edition

Although it is only two years since the second edition of this HANDBOOK was published, the continuing demand for it has made possible another edition. Clearly, it has proved to be a useful little book and this new edition has enabled a thorough revision to be undertaken. The geographical scene is interestingly, but annoyingly, dynamic and many of the statements, which were acceptable and also correct only a few years ago, are now no longer true; hence, an attempt has been made to eliminate such inaccuracies. This new edition allows some restructuring of the text to take place and further elaborations to be made: for example a new chapter on the early steps in human progress has been added, Chapter II has been expanded to introduce some basic concepts of demography, and an additional section on the political geography of the oceans has been put into Chapter IX. Also, some new and additional tables have been introduced. It is hoped that these improvements will further enhance the value of the book and make it a useful text for the geography content of TEC/BEC courses.

August 1978   H. R.

Contents

List of Illustrations

CHAPTER I

The Scope and Purpose of Human Geography

THE STUDY OF GEOGRAPHY

1. Early geography. The word "geography" is derived from the Greek words *ge*, meaning the earth, and *graphe*, meaning a description. Hence, geography may be defined, in its simplest way, as the description of the earth.

In earlier times geography was purely descriptive and concerned itself with listing facts about places and peoples; this has sometimes been dubbed gazetteer geography. Although some facts are of interest in themselves and although it is necessary to know many facts in modern geography, factual knowledge is rather arid. Certainly the old-fashioned geography with its long catalogues of names of capes and bays and towns and commodities was not a particularly interesting or inspiring subject. The early maps, too, were mainly concerned with places, the locating of mountains, rivers and towns, and if the map-makers did not know the geographical facts about a particular area, they were prone to use their imaginations to fill the empty spaces.

2. The influence of the discoveries. The geographical discoveries of the fifteenth and sixteenth centuries and the exploration of the continents which followed had important effects upon the growth of geography. The improvements in navigational instruments and the interest in mathematical geography led to improved and more accurate cartography. The opening up of new lands and the acquaintance of foreign peoples brought a new interest in faraway places and introduced people to a variety of new commodities, e.g. maize, potatoes, tobacco, cacao.

The early development of geography was largely concerned with the collection, systematisation and correlation of the vast wealth of new information which discovery and exploration had placed at the geographer's disposal. The vast accumulation of facts also resulted in geography narrowing its field of enquiry; it began to shed some of those aspects of terrestrial phenomena which more properly belong to other sciences, such as physics, geology and biology.

In the nineteenth century a number of scientific explorers and travellers, who were also great naturalists, added much to the knowledge and understanding of the distribution and nature of physical and human phenomena. Among these great naturalists were the following:

(*a*) Alexander von Humboldt, a German, who travelled widely in South America and Siberia.

(*b*) Sir Joseph Hooker, who travelled in northern India and visited Antarctica.

(*c*) A. R. Wallace, who carried out important work in biology in south-east Asia.

(*d*) Charles Darwin, who voyaged around the world in the *Beagle.*

3. The beginnings of scientific geography. About a century ago Charles Darwin published his well-known book, *The Origin of Species,* in which he suggested that all forms of life had evolved from earlier, more primitive forms. At that particular time this revolutionary idea caused a great stir in scientific circles. Not only did it affect the sciences of biology and anthropology but it affected almost every other science: it set people re-thinking, questioning and experimenting.

In the realm of geography, for example, people were no longer content to learn that kangaroos lived only in Australia or that the Dead Sea was extremely salty; they wanted to know why kangaroos were to be found only in Australia and why the Dead Sea was so much more salty than other seas. This inquisitiveness led eventually to the growth of causal geography: geographers began to seek for the causes of things and, also, to note the effects of conditions and happenings upon the earth and man. This gave rise to what is termed "cause and effect", and geographers are still very much concerned with causes and with effects.

Modern geography, therefore, not only describes the surface of the earth and the various features, both natural and man-made, upon it, but it also investigates and correlates, organises and rationalises the phenomena of the earth. It is a descriptive and analytical study: it has, moreover, its own particular point of view; hence, geography has been called at once an art, a science, and a philosophy.

4. The scope of geography. The field of geographical study is very wide: some people believe it is too wide but this is an arguable matter. Amongst the many subjects with which the geographer is concerned are the following:

(a) The size, shape and movements of the earth.

(b) The distribution and position of land-masses and bodies of water.

(c) The rocks, structure and relief of the earth's surface.

(d) The waters of oceans, seas and lakes and their movements.

(e) The atmospheric conditions and the varying climates to which they give rise.

(f) The earth's vegetation and the distribution of animals.

(g) The races of mankind and the distribution of population.

(h) The activities of man and the ways in which he makes a livelihood.

(i) The different kinds of settlement in which man lives.

(j) The social and cultural characteristics of human groups.

(k) The political organisation and relationships of human groups.

All these topics, physical and human, the geographer carefully studies, examining their spatial distribution and noting their mutual relationships. Briefly, geography may be said to study the "where", "why" and "how" of things.

5. The geographical viewpoint. Before we go on to elaborate upon the human aspects of geography it will be useful to indicate briefly the principal concepts of the geographer. The geographer's task is to explore, understand and interpret the physical and cultural surface features of the earth and in doing this he views the earth with four particular concepts constantly in mind.

(a) The concept of spatial relationships. Everything that happens occurs in time and space. All the features of the environment, whether natural (physical) or man-made (cultural), have evolved through time; hence the geographer is especially interested in the processes by which the present-day features or elements of the environment or habitat have come into being and in the variations of these features from place to place. He is very much concerned with what is usually termed the spatial-locational aspect and he views the elements of the landscape, and the processes by which they came into being, in terms of the areas or spaces which they occupy and their locations. Expressed differently, and perhaps more simply, the geographer is concerned with distributional relationships and areal differences.

(b) *The concept of man–land relationships.* The geographer explores and analyses the interrelationships which exist between human society and the natural environment. Inevitably, by the very nature of things, there is bound to be a strong relationship between man and his habitat. The environment influences, to a greater or lesser degree, man's activities, settlements and culture: but the interrelationship between man and his habitat is a reciprocal one; hence man not only adjusts himself to his surroundings, but sometimes overcomes and changes them. However, whether nature or man is dominant, there is a constant interaction and interdependence between them.

(c) *The concept of man–man relationships.* Half a century or so ago the geographer sought explanations for man's distribution, behaviour and activities largely in terms of the physical environment and this gave rise to what came to be known as geographical determinism. The modern geographer's search for explanation has caused him to move away from the earth sciences towards the social sciences, for there is a growing appreciation that many relationships are of a man–man kind rather than a man–land nature. The geographical landscape is increasingly being changed and fashioned by human decisions and it is now realised that the explanation of it results as much from behavioural (human) and stochastic (chance) influences as upon physical and economic forces: think, for example, of local government reform, of the new town, of the inner city problems, of the immigration problem, of Scottish and Welsh nationalism, in Britain.

(d) *The concept of environmental unity.* The environment is composed of many heterogeneous elements: terrain, soil, climate, vegetation, animal life, people, settlements, etc.; between these there is always some interplay, some relationship. These elements are never separate, discrete or unrelated. Forces and processes are continually at play, linking and uniting the various elements into a whole, into an integrated, coherent entity. Hence we arrive at the conclusion that the habitat is the sum of all the environmental elements, forces and processes (physical and human) interacting with each other. Out of this idea of environmental unity arose the regional concept, i.e. that unit-areas of the earth's surface can be differentiated because they possess specific, identifiable characteristics which, at the same time, give unit-areas or regions a recognisable degree of unity, synthesis and geographical personality.

THE BRANCHES OF HUMAN GEOGRAPHY

6. Human geography. We have already suggested that there are two principal aspects to geography, physical and human; and for convenience we refer to the branches of physical geography and human geography. But it would be very foolish to think of these two branches as being quite distinct, unrelated and separate areas of study; they are, on the contrary, most closely related and interact one with the other. As we have just said above, the geographer believes in environmental unity.

Accepting this division for convenience, however, let us now try to define the scope and subject-matter of human geography. Many years ago the late Professor P. M. Roxby gave a definition of human geography; he wrote that it "consists of the study of (a) the adjustment of human groups to their physical environment, including the analysis of their regional experience, and of (b) inter-regional relations as conditioned by the several adjustments and geographical orientation of the groups living within the respective regions." It is interesting to note that a modern geographer, Professor Emrys Jones, echoes Roxby's words; Jones writes that human geography is concerned with "those aspects of human life which, through a continual and changing interaction with nature in all its forms, have given rise to distinctive landscapes and regions". (*Human Geography*, Chatto and Windus.)

Following these leads we may define geography as the influence which the environment exercises upon the life, activities, progress and distribution of man but recognising, at the same time, the reciprocal action of man upon his environment, an action which becomes increasingly intensified and powerful as man's scientific knowledge and technological skill advance .

7. The scope of human geography. The scope of human geography has had different interpretations. The French geographer Jean Brunhes took a somewhat limited view of the field of human geography. He maintained that human geography should be confined solely to those observable features impressed upon the earth's surface as a result of man's activities. Believing emphatically in the physical basis of all geography, Brunhes divided the human aspect into three parts each of which was based upon fundamental facts:

(*a*) *Facts of unproductive occupation*, e.g. houses, settlements, roads.

(*b*) *Facts of plant and animal conquests*, e.g. cultivated fields, domesticated animals.

(*c*) *Facts of destructive economy*, e.g. the exploitation of minerals, animal and plant devastation.

Brunhes maintained that these essential facts covered all the material of human geography and he excluded everything else.

Roxby distinguished four principal aspects: racial, social, economic and political. Brunhes excluded the racial and political, as well as certain social aspects from the true content of human geography, since, he maintained, the human problems connected with these aspects could not be solved by geography alone. The argument for their inclusion in human geography is that since the core of human geography is the relation of man to his habitat then such aspects of his life as the racial, social and political aspects are important and ought to be studied.

During more recent times it has become customary to exclude the racial aspect as a distinctive area of study from the content of human geography since it was generally felt that this belongs more properly to the field of anthropology. When it is studied, it is in connection with population studies which form part of the social aspect. Some modern geographers challenge the right of political geography to be a part of human geography but it is of interest to note that a book on human geography written by an authority on the subject, A. V. Perpillou, excludes any consideration of the racial aspects but does include, albeit rather briefly, a chapter on the political aspect. Since Perpillou follows the tradition of French human geographers his inclusion of the political aspect is rather surprising but we must interpret this as meaning that he approves of political geography being considered as a branch of human geography.

8. The sub-divisions of human geography. Without overlooking the essential unity of human geography, it is common practice, if only for convenience of practical study, to sub-divide human geography. (For the interrelationship of all these subjects, *see* Fig. 1.) Three main divisions may be recognised:

(*a*) *Social geography* which deals with the growth and distribution of population, settlement types and their distribution and human cultural features such as religion, language, community organisation, etc.

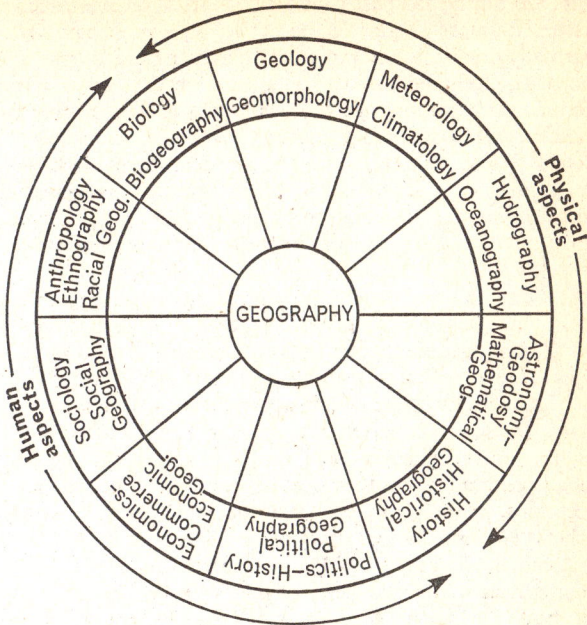

FIG. 1 *Wheel graph of geography.*

The graph illustrates the main relationships which exist between geography and the other arts and sciences. Geography may be regarded as a bridging and integrating subject.

(b) *Economic geography* which embraces the study of the exploitation of natural resources, the production of commodities, the location and distribution of manufacturing industries and international trade and communications.

(c) *Political geography* which is concerned with political units, their territorial areas, boundaries and capitals, with the elements of national power and with international politics which are considered from the geographical point of view.

9. **Recent changes in emphasis.** From what has been said so far it will be clear that geography has changed in its approach and content over the centuries. But perhaps the greatest of all has

happened during the past twenty-five years, a change which has amounted to a revolution in the subject. Pre-war geography was essentially descriptive and qualitative in its nature but the adoption of scientific method and quantification has radically transformed the subject and modern geography is emphatically more analytical and explanatory and has become more problem orientated. Fundamentally, it may be said that the modern revolution in geography lies in its conceptual basis and its increasingly theoretical nature. Although this makes modern geography more difficult, it also makes it more interesting and more scientifically respectable.

At the same time there have been changes in the foci of interests in the field of human geography. One of the earliest interests was in urban geography, an interest which still looms large and engages the attention of many geographers. Another early interest, particularly with American geographers, lay in cultural geography, though this never became very highly developed in Britain. American geographers, too, were pioneers in the area of medical geography, another facet of geography which, with a few exceptions, never really engaged the interest of British geographers. Perhaps the two areas of social geography which have most caught the interest of British geographers are recreational geography and man's concern for man which has come to manifest itself in what the American's call the "geography of care". Likewise in the fields of economic and political geography there have been changes in emphasis with new areas of interest sprouting up, e.g. marketing geography and the political geography of the oceans.

PROGRESS TEST 1

1. Explain the terms "gazetteer geography" and "causal geography". (1, 3)
2. Explain how the new scientific ideas of the nineteenth century influenced the development of geography. (3)
3. Describe the scope and content of geographical study. (4)
4. Geography has a distinctive viewpoint. Which particular concepts does the geographer have constantly in mind? (5)
5. Discuss the scope of human geography. (6, 7)
6. Compare Brunhes' and Roxby's interpretations of the field of human geography. (6, 7)
7. Look up the meaning of anthropology in a dictionary or

encyclopedia. In terms of this meaning do you think that racial geography is a justifiable branch of human geography? (7)

8. Which are the main branches of human geography? Outline the content of each branch. (8)

9. State what is meant by the modern revolution in geography and indicate some of the new areas of interest which have emerged in the field of human geography. (9)

CHAPTER II

Race and Population

THE RACES OF MANKIND

1. The meaning of race. "Race" is a word which gives rise to much confusion. Many people associate race, erroneously, with such factors as language, religion and nationality. These, however, are merely cultural veneers and are quite superficial, for anyone, if he so wishes, may change them. On the other hand, racial character cannot be changed, for it is fixed for us at conception and we are born with it. In other words, race is simply biological inheritance.

Race refers to physical characteristics, such as skin colour, stature, physique, hair type, etc., physical traits as the scientist calls them, which are passed on through the blood from parent to offspring. Whether mental characteristics are similarly passed on has been disputed, but it seems very likely that this is so.

2. Origin and dispersal of man. Man has a very long ancestry of several million years but all types of modern man are descended from the species *Homo sapiens* (wise man). Before *Homo sapiens* there were many other types of "man", all descended from ape-like creatures, though these types all became extinct. Only those creatures capable of pitting their wits against nature managed to survive; they formed *Homo sapiens*. The old saying that man is descended from monkeys is not true; but men and monkeys do come from the same remote ancestral stem.

Where the ancestral type of modern man first emerged is not known. At one time it was commonly assumed that central Asia formed the original home of man, but recent discoveries in Africa make that continent a more likely possibility. There is fairly general agreement, however, that man originated in tropical or sub-tropical lands, but not in either Europe or the Americas. Wherever the original centre of dispersal was, the fact remains that early man wandered, perhaps quite unconsciously, in various directions until eventually he came to inhabit most of the earth's surface.

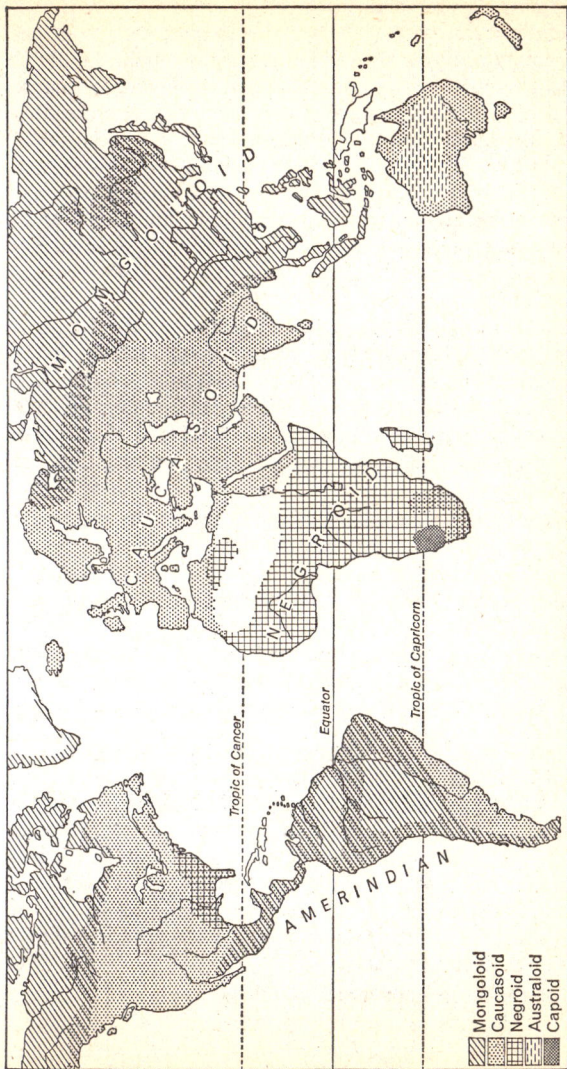

FIG. 2 *The races of the world.*

This is a simplified map, showing the distribution of the principal ethnic groups in the world. In some parts of the world, notably in Central and South America and south-eastern Asia, there has been much racial intermixture. Scattered pygmy groups (Negrillos and Netritos) are also to be found in Central Africa and south-eastern Asia.

Mongoloid
Caucasoid
Negroid
Australoid
Capoid

3. Racial purity. Man is a mobile creature and is the only animal to be found in every part of the world, and these facts, together with the ability of any male and female of the human family to mate successfully, have led to widespread interminglings; because of this, there is no such thing as a pure race. The interbreeding that has taken place over many millennia between the original racial groups has blurred all racial boundaries. For this reason, many scientists have advocated the use of the term ethnic groups (ethnic means pertaining to race) rather than racial groups to describe broad groups of people having certain physical characteristics in common. Figure 2 shows the distribution of the principal ethnic groups in the world.

Possibly, with one or two exceptions, such as the Kalahari Bushmen and the inhabitants of the Andaman Islands in the Indian Ocean, there are no peoples of racial purity. While it is inadmissible to speak of pure races, it is true that segregation and isolation tend to give rise to more or less fixed types; hence it is possible to speak of relative ethnic purity. For example, there are these two groups:

(a) The people of Scandinavia form an especially distinctive group, i.e. fair-haired, fair-skinned people, which may be explained in terms of the area's relative isolation and the fact that, historically, people have moved out of, rather than into, Scandinavia.

(b) The aborigines of Australia, the Blackfellows, were cut off in this rather remote and isolated continent for so long that as a group they remain physically distinctive and relatively pure.

4. How are ethnic groups distinguished? Anthropologists first divided the human race on a basis of skin colour, the most obvious differentiating physical characteristic. They recognised five groups: white, yellow, red, brown and black. Although colour of skin provides an easy method of separating one group from another, the use of skin colour proved to be neither a satisfactory nor scientific way of grouping the different peoples of the earth. How unsatisfactory skin colour is as a criterion of race is shown by the following facts:

(a) Both the Negroes of Africa and the aborigines of Australia have very dark skins yet they belong to two quite distinct ethnic groups.

(b) Most of the brown-skinned peoples of the world belong to the same "race" as the white-skinned Europeans.

Skin colour is far from being a reliable guide, hence anthropologists use additional physical characteristics, e.g. shape of the head, stature and body proportions, shape and form of the nose, to help them group mankind.

5. Chief criteria. Of the many physical traits which can be used by the anthropologist to differentiate people, head form and hair are, perhaps, the two most important.

(*a*) *Cranial form*. The ratio between the length of the skull and the breadth is known as the cephalic index. Measurements have shown that some people have long and narrow heads (dolicocephalics), others broad and round heads (brachycephalics), while other have heads of an intermediate index (mesocephalics).

(*b*) *Hair*. This differs in colour (fair, brown, black), in texture (lank, wavy, frizzy) and in cross-section (round, oval, flat). Classification by hair gives three main groups: the straight-haired peoples, the curly- or wavy-haired peoples, and the woolly-haired peoples.

NOTE: Modern research has shown that there are possibilities of using blood groups as a basis for classification.

6. Classification of races. Anthropologists differ among themselves as to exactly how many distinct ethnic groups there are, but most recognise five or six main types. The distinguished American anthropologist, Dr. Carlton S. Coon, has recognised five primary groups:

(*a*) The *Caucasoid* (white): Europeans and peoples of European ancestry and brown-skinned people such as the Arabs and Hindus.

(*b*) The *Mongoloid* (yellow): Mongols, Chinese, Eskimos and the American Indians.

(*c*) The *Negroid* (black): the Negroes of Africa and America.

(*d*) The *Australoid* (black): the Australian aborigines.

(*e*) The *Capoid* (brownish yellow): the Bushmen and Hottentots of South Africa.

Certain pygmy peoples, the Negritos and Negrillos, scattered in small groups throughout the tropical belt, probably form a sixth group.

The peoples of Indonesia and Polynesia create something of a racial problem but it would seem that they are of mixed origin with a basic stratum of Mongoloid race in their ethnic make-up.

7. Ethnic numbers. The Caucasoids form the largest single group and account for nearly half the world's people. They are found in a wide zone running obliquely from north-west to south-east across the Old World land mass and also in the "newer" lands of the world, i.e. North and South America and Australia. The white- and brown-skinned peoples are closely followed by the Mongoloids, whose numbers continue to grow rapidly. The Negroid peoples, whose numbers remained stationary for several centuries, have recently—mainly in the present century—begun to increase at a fairly rapid rate. Their total world number is, however, small in comparison to the Caucasoid and Mongoloid totals.

Whereas these three groups are all flourishing and expanding rapidly in their numbers, the pygmies, aborigines, Bushmen and Hottentots, already numerically very small, are dying out. It is very difficult to foresee the future but it is possible that one hundred years from now they will virtually have disappeared as distinctive groups, for they are likely to have died out or become assimilated by other groups.

8. Racism. Racism is difficult to explain but it is an acute problem in the world today. Some people believe it is an instinctive defence mechanism; others that it has its roots in civilisation and culture. Certainly it is not something that is new, for it has been known in the Indian sub-continent for at least two millennia; indeed, the caste system was partially racial in its origins. Racism, however, would seem to be closely associated with colour, and the darker a man's colour the more virulent usually is the racial antipathy.

9. Melting-pot of races. One of the most interesting human experiments going on in the world today is in Latin America. Here, racial intermixture has been taking place for several centuries on a large scale and interbreeding between the three racial groups (whites, Negroes and Amerinds) has given rise to several hybrid types. These hybrids, in turn, are intermixing with the result that the entire region, with few exceptions, is a great melting-pot of races. In due course a new ethnic "type" is likely to emerge. In Latin America colour consciousness has seldom manifested itself and unless racism is deliberately aroused there this ethnic experiment should be successful and perhaps teach the rest of the world a lesson. Lest we may be inclined to prejudge the issue, it would be well for us to remember that the people of Britain are of very mixed ancestry and the historian H. A. L. Fisher aptly called the British a group of "energetic mongrels".

THE PRINCIPLES OF DEMOGRAPHY

10. The study of population. The study of population is termed demography, and it is important for the human geographer to know, for any country or area he may be studying, the following facts:

(*a*) The size or numbers of the population.
(*b*) The density and distribution of the population.
(*c*) The growth during recent times.
(*d*) The current rate of growth and movements.
(*e*) The age structure of the population.
(*f*) The balance between males and females.
(*g*) The standard of literacy.

These are matters of considerable consequence to the state government for they have economic and social and even politico-military implications.

11. Population dynamics. Something must be said, if only briefly, about the dynamics of population growth: in other words, about the significance of such factors as birth rates, death rates, nuptiality, fertility rates and the like, for these strongly influence, in particular, growth rates and population totals.

The total population of an area or country is the balance between two sets of forces: the natural change and migration change components. The natural change or reproductive change is the difference between fertility and mortality, i.e. births and deaths. It will be obvious that if births exceed deaths in a given period, the population total will increase; but if the opposite occurs, then a decline in the population will result. The rate of increase of a given population may be expressed by the following formula:

$$r = b - d$$

This simple relationship is modified, however, by the migration change component, for demographic gains or losses may occur through immigration or emigration; if immigrants outnumber emigrants the population will increase, and vice versa.

It will be clear that population totals are the result of four factors: births and immigrants which increase the total, and deaths and emigrants which lower the total; there are, in fact, only two possible ways by which a population can change: through a natural rate of increase and through net migration.

Birth rates are expressed as births per 1,000 people per year: they are arrived at by dividing the total number of births by the estimated population at the middle of the year and multiplying the result by a thousand to give the figure per 1,000 people Such a figure is termed the crude birth rate. The term "crude" is used because when birth rates and death rates for different countries are compared on the basis of such figures the age structure of the population is not taken into consideration. Sometimes the term net reproductive rate is used; this is a more refined way of expressing birth rate and refers to the number of births per female of a specific age class. The fertility rate, which is a modification of the net reproductive rate, refers to the number of births per 1,000 women of child-bearing age, i.e. 15 to 44 years of age.

12. Ecological periods. Demographers frequently divide human populations into three ecological periods or age groups as follows:

(a) The pre-reproductive group, which corresponds, in general terms, to the dependent young, ages 0 to 14.

(b) The reproductive and economically active group between the ages of 15 to 65.

(c) The post reproductive group, i.e. elderly dependants over 65 years of age.

Within any population group these ratios may show considerable variations.

A necessary exercise for the demographer is his consideration of the age structure of a population since this enables him to assess its growth potential. This he does by means of age pyramids which illustrate the age composition of the population (*see* Fig. 3).

Such age population pyramids are constructed by using bar graphs, one set representing males of different ages, the other females: these are drawn back to back against a vertical line which marks zero. The pyramids can be drawn to show absolute numbers in different age groups or to show the proportion of the total population falling within each of the age groups.

Differences in the shapes of population pyramids permit:

(a) the population structures of two different countries to be compared;

(b) the structural changes in the population of a particular country to be compared over a given time span.

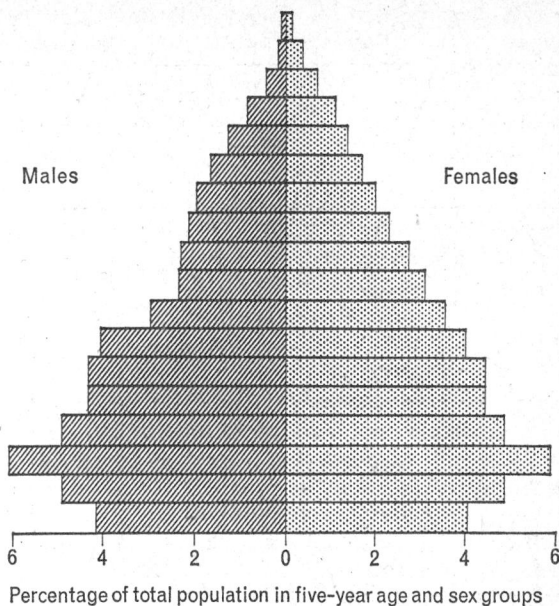

Males   Females

6   4   2   0   2   4   6

Percentage of total population in five-year age and sex groups

FIG. 3 *An age population pyramid.*

The shape of a population pyramid can reveal much about the present and future of a population.

13. Population pyramids. Although no two population pyramids are ever likely to be precisely alike, five main types of pyramid may be distinguished as illustrated in Figure 4.

(*a*) A regular triangular-shaped pyramid is characteristic of countries having high birth rates and high death rates. It indicates a youthful population with a small old-age group and foretells a rapid growth in population as the young people enter into the reproductive period. Such pyramids were typical of most countries up to the eighteenth century, but relatively few fall into this category at the present time. Mauritius and some of the countries of Latin America may be said to have a population structure resembling this shape.

(*b*) This pyramid with its concave sides is characteristic of most

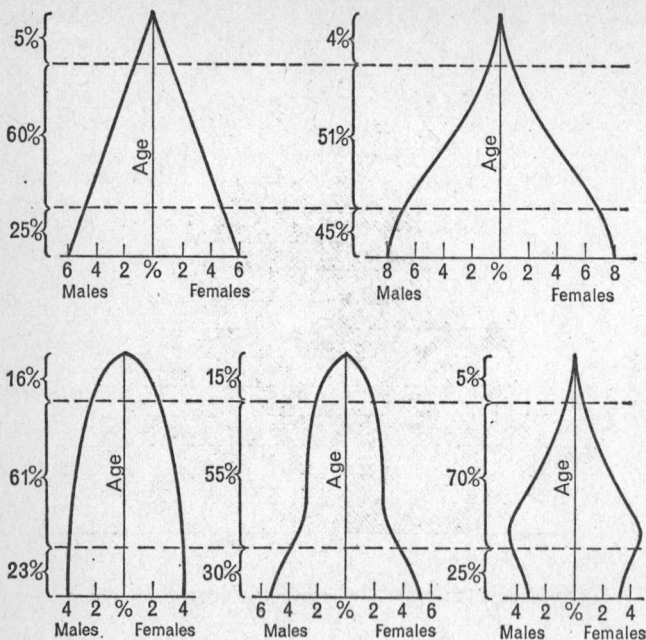

[*From* Population Problems *by D. T. Lewis and W. S. Thompson. Copyright McGraw-Hill, 1965. Used with permission of McGraw-Hill Book Company.*]

FIG. 4   *Types of population pyramids.*

of the developing countries, e.g. Mexico, India, Sri Lanka. This pinched pyramid indicates a population structure in which death rates are beginning to decline, especially in the 0 to 5 age group, and in which birth rates are still high. The shape is indicative of a rapidly expanding future population.

(*c*) A narrow-based pyramid with more or less convex sides denotes countries with a low population growth and low death rates. The ratio between one age class and another varies little. Such pyramids are indicative of no significant population growth and a tendency towards an aging population. Britain and Sweden are typical of this type, but most western European countries fall into this category.

(*d*) A bell-shaped pyramid indicates that there is a relatively large ratio of the population on the pre- and post reproductive age periods. The marked narrowing in the reproductive age period indicates a drop in the birth rate at an earlier time. The structure shows that there has been a renewed increase in the birth rate following a period of low birth and death rates. The shape is characteristic of some of the developed countries, e.g. the United States and Canada.

(e) This pyramid shows a population structure in which death rates are declining but, more importantly, where there has been a marked drop in fertility. This condition was typical of Britain and France in the period between the two World Wars. At present Japan is the only major country with a pyramid shape of this kind, but the current trend in the birth rate in Britain strongly suggests that within the next decade it, too, will develop this shape.

THE GROWTH OF POPULATION

14. The growth of population. It is possible that around 5000 B.C. the total world population was only of the order of some 10 millions. By the time of Christ it has been estimated that the earth's peoples totalled about 100 millions. By A.D. 1600 world population was probably around 400 millions.

Population growth was slow, and may even have oscillated slightly, until the beginning of modern times. A number of reasons help to explain this slow expansion:

(a) *Natural disaster*, such as earthquake, fire and flood, against which man had little defence, took a heavy toll on the population.

(b) *Famines* in earlier times were of frequent occurrence in many areas and no doubt periodically decimated the population.

(c) *Pestilence and disease* often ravaged the population, e.g. the Black Death in the fourteenth century, which in England mortally affected 1½ million people or between one-third and one-half of the population.

(d) *Wars* had the effect of drastically restraining population growth.

(e) *Persecution*, whether of a political, religious or other nature, caused many deaths and so helped to keep down the population.

(f) *Social habits and customs*, such as the practices of human sacrifices, head-hunting and suttee (the practice of Hindu widows immolating themselves on the funeral pyres of their husbands).

15. The revolution in population growth. Although the world's population showed a steady increase over a period of several thousand years, the growth was fairly slow. By 1750, that is on the eve of the Industrial Revolution, the total had grown to around 800 millions. Then, suddenly, after 1750, the numbers began to grow very rapidly and during the next 200 years they increased almost fivefold (*see* Fig. 5).

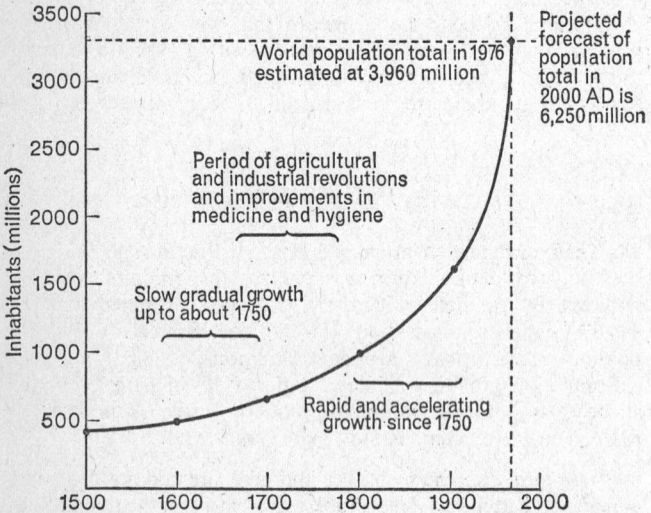

FIG. 5 *Graph of world population growth.*

Note how the population remained fairly stable during the sixteenth century, how it began to increase more rapidly during the period 1750–1850 (the time of the Agricultural and Industrial Revolutions) and how since about 1850, population growth has accelerated.

It is usually said that the Agricultural and Industrial Revolutions were mainly responsible for the acceleration of population growth. No doubt there is much truth in this statement but it may be argued that it was the growth in population that helped to stimulate these Revolutions. About this time two factors do seem to have been largely responsible for the population expansion:

(*a*) Improved food production and the increased variety of foods.

(*b*) Increasing medical knowledge and improvements in hygiene.

16. The population explosion. The improvements in health and hygiene and food supply reduced the effects of the disasters brought by plagues, diseases and famine; the developments in science and technology brought increasing control over natural disasters, while increased law and order in many parts of the world helped to reduce tribal and civil wars. The cumulative effect of these various factors resulted in a very rapid expansion in the total number of people.

At the present time the population in the world is increasing at the rate of about 12,000 per hour or 300,000 per day or over 100 millions per year. At the current rate of growth there will be at least 4,200 million by 1980 and some 6,250 by 2000 A.D. This expansion has been so great during recent generations that we have come to speak of a "population explosion" (*see* V).

17. Differential growth. The reader should note that the population is not increasing at the same pace everywhere. While the world growth is at the average rate of 1.9 per cent per annum not all countries show such an increase: Eire and Norway, for example, show a lower rate of increase, while, on the other hand, some countries such as Mexico and Kenya show a rate of increase of over 3 per cent. Again, we must beware of assuming that the countries with the largest populations, such as India, have the greatest rates of increase. This is not always the case: growth rates in Latin America (which has a total population of only about 300 millions) are much higher, often twice as high as those in Monsoon Asia (which has over 2,100 millions). However, it is true to say that the aspect of the rise in population which is causing the greatest concern is the rapidly increasing totals of the already numerically large countries of southern and eastern Asia.

POPULATION DISTRIBUTION

18. World distribution. If we examine a world map showing the distribution of population, such as the simplified map in Figure 6, we are immediately aware of the irregularity of this distribution: some areas are very densely peopled, others very thinly

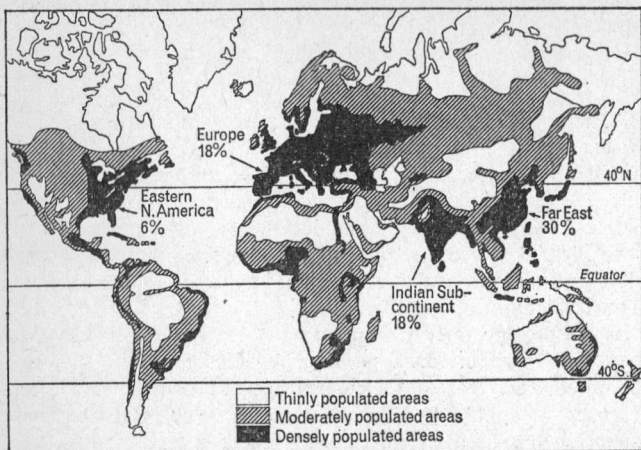

FIG. 6   *World distribution of population.*

The world's population is very unevenly distributed. Note the four areas of dense population, all of which are in the northern hemisphere and lie mostly in temperate or sub-tropical latitudes. Nearly three-quarters of the world's people live in these four areas. Note, also, the location of the thinly peopled areas, i.e. the ice-cap, tundra, forest, desert and high altitude areas.

peopled. Even areas in close proximity, e.g. as in the Near East, the Indonesian islands, South Africa, show greatly differing densities.

The map, however, demonstrates quite clearly that mankind is massed in four main regions of the earth:

(*a*) In western and central Europe, especially in Britain, France, the Low Countries, Western Germany and Italy.

(*b*) In the east central part of North America, i.e. the eastern United States and south-eastern Canada.

(*c*) In the Indian sub-continent comprising Pakistan, India, Bangladesh and Sri Lanka.

(*d*) In the far east of Asia, especially in eastern China, Korea, Japan, Hong Kong and Taiwan.

Outside these main areas of population concentration, mankind is much more thinly spread, though here and there are clusters of fairly dense or very dense population, e.g. in Egypt, Nigeria, Java, coastal New South Wales, La Plata estuary, south-eastern Brazil, California.

19. Thinly peopled areas. There are many areas, often of great size, which are very thinly peopled. These are usually areas of in-accessibility, high relief or unfavourable climate. Typically, the density is under 0.8 km² and, often, under 0.4 km². The follow-ing are the chief areas in this category.

(*a*) The cold tundra and ice-cap regions of polar lands, i.e. Canada, Greenland, Siberia, Antarctica.

(*b*) High altitude lands such as the Himalayan country and the high, wind-swept plateaux of Central Asia and portions of the cordilleran system of the Americas.

(*c*) The hot, arid areas which are devoid of water supplies such as most of the Sahara, the Arabian Desert and the Australian Desert.

(*d*) Areas of hot, wet forest such as the Amazonian selvas and forested Borneo, Papua, New Guinea and parts of the Zaïre basin.

20. Local variations. One of the interesting facts about popula-tion distribution is that even in an area which as a whole is shown to have a high density, such as India or the eastern U.S.A., there are notable variations in density between one area and another. Similarly, areas which are lightly peopled may have pockets of high density. Only a detailed population map will indicate these variations.

Two questions arise: Why does population density vary so much over the earth's surface as a whole? Why does population density vary so much between place and place even in a relatively small area? We have already hinted at some of the answers but let us elaborate further.

21. Factors affecting population distribution. The population of any area is at once the cause and the result of its economic possibilities. Man usually congregates where natural conditions most easily offer a supply of food or provide a means of earning a livelihood. Where conditions are difficult, i.e. where the chances of finding a food supply are limited or the opportunities of earn-ing a living are closely restricted, man is not tempted to settle down and reproduce his kind.

The chief geographical factors influencing the distribution of population are as follows:

(a) Accessibility.

(b) Surface relief (altitude and the character of the topography).

(c) Climatic conditions.

(d) Natural vegetation.

(e) Soil fertility.

(f) Water supplies.

(g) Mineral deposits.

The bulk of the world's population, some 70 per cent, live by agriculture, hence the conditions of relief, climate and soil are of fundamental importance since they largely determine the food supply. Very mountainous country makes crop growing difficult, if not impossible. Again, if it is too cold or too dry or the growing season is too short, crops will not grow. Also successful agriculture needs good soils; some soils are so sterile that crops just cannot be grown in them. It is no accident that the great lowlands, which are reasonably level and usually have deep, fertile soils, such as alluvium, are mostly areas of dense population. Conversely, high, rugged, mountainous areas with their thin, stony, immature soils and cold, windy climates are very sparsely peopled.

Water is essential for agriculture, industry and settlement and its presence or absence has greatly influenced population distribution. The presence of mineral wealth is often an attractive force, especially in this technological age, and some of the greatest concentrations of population are to be found on some of the coalfields and other mineral fields: for instance, the Ruhr Coalfield in West Germany and the Appalachian Coalfield in the United States. The occurrence of valuable mineral deposits, such as gold, copper, tin and uranium, has caused man to settle in the most unattractive spots such as the copper mining centre of Chuquicamata in northern Chile which lies at over 3,000 m in the Andes, or Uranium City in the icy wastes of northern Canada.

Finally geographical position, in relation both to means of transport and to accessibility, has been an important factor influencing population distribution. The fact that the eastern part of North America is more densely populated than the western is partly due, though only partly, to its being nearer to western Europe whence the early settlers came. Again, the fact that the island continent of Australia was rather isolated and remote explains to a very large extent the lateness of its settlement by Europeans and helps to account for its still relatively small population. The same applies to New Zealand.

22. Non-geographical factors. In addition to physical geographical factors certain human factors, e.g. of an historical and political kind, have sometimes influenced settlement of areas and population densities within areas. One can think of the settlement of New England by the Pilgrim Fathers, of Salt Lake City by the Mormons, of Filadelfia in Paraguay by the Mennonites or of the enforced drafting of peoples into Siberia to open up that land or, again, of the deliberate dispersion, sometimes liquidation, of peoples by governments in the past. It is interesting to recall that the first real settlement in Australia began with the establishment of a penal colony.

23. Density. It will be clear that the density of population in any area is a fairly complex matter and results from a combination of factors both geographical and non-geographical. The earth's land surface is often divided into areas having high, moderate and low densities of population.

(a) *High density areas* may be sub-divided into two:

(i) Those having high living standards, e.g. England, Netherlands, France, West Germany, north-eastern United States.

(ii) Those having low living standards, e.g. India, Java, the Great Plain of North China.

(b) *Moderate density areas* may be sub-divided into two categories:

(i) Those with high living standards, e.g. Denmark, New South Wales, the St. Lawrence lowlands of Canada and much of the Mid-west of the United States.

(ii) Those having low living standards, e.g. Turkey, Cambodia, Portugal, and much of eastern Brazil.

(c) *Low density areas* may be divided into the following:

(i) Those with high living standards, e.g. Sweden, New Zealand, British Columbia.

(ii) Those with low living standards, e.g. Paraguay, Tibet, Mali, Borneo.

MOVEMENTS OF PEOPLE

24. Motives. An important aspect of demography is the movement of people from place to place. Movements are of three main kinds:

    (*a*) *Unconscious drifts,* such as the wanderings of early man.

    (*b*) *Enforced movements,* such as the transportation of slaves.

    (*c*) *Voluntary migrations,* such as the modern migrations to Commonwealth countries.

There has always been *some* movement of people throughout history but this has often been fitful. From time to time, there have been large-scale movements as, for instance, the case of the barbarians in the third to the fifth centuries, the Mongols in the thirteenth and fourteenth centuries, the western Europeans in the sixteenth and seventeenth centuries and the Bantus in the nineteenth century.

In the beginning man's movements were unconscious wanderings and, as a result of these driftings, he came to occupy most parts of the earth's surface. Later in history man moved consciously and mainly for two types of reasons which we might term "attractive" and "repellent" reasons (*see* **25** and **26** below).

25. Attractive reasons. Among the varied reasons which attracted man to move and settle elsewhere were the following:

    (*a*) More congenial climatic conditions.

    (*b*) Better economic opportunities.

    (*c*) Higher standards of living.

    (*d*) Freedom of thought and belief.

    (*e*) Adventure and the sheer attraction of "foreign parts".

Thus many Irishmen, Bangladeshis and West Indians have come to England because this country offered them better economic opportunities. Many British subjects have emigrated to Australia and New Zealand which promise (but do not always fulfil) better conditions of life and livelihood. Note in this connection the "brain drain" to the United States.

26 Repellent reasons. Some people quit their homelands, voluntarily or involuntarily, for these reasons:

    (*a*) Harsh environmental conditions which make getting a living precarious, e.g. in Highland Scotland, Norway.

    (*b*) Natural disasters such as earthquakes, volcanic eruptions, floods and famines.

(c) Religious persecution, e.g. the Huguenots who left France after the Revocation of the Edict of Nantes in the reign of Louis XIV and the Puritans in seventeenth-century England.

(d) Political persecution, e.g. the Jewish refugees from Nazi Germany and the Baltic peoples from Russian communism.

(e) Economic depression and lack of work which compel people to seek work elsewhere.

These and other reasons have made people move. One can think of the 60 million people who left Europe between 1820 and 1930 to try their luck in the "new lands" of the world.

27. Economic motives. These loom large in all human movements. Examples are as follows:

(a) *The Viking outpouring* from the Scandinavian lands during the Dark Ages was ascribed by H. A. L. Fisher to "common cupidity": they went in search of loot and land and trade.

(b) *The Portuguese and Spaniards* who went to the Americas in the sixteenth century were motivated by the promise of "get rich quick", though few in fact found the legendary stores of gold, silver and precious stones.

(c) *The southern Chinese* who migrated into south-east Asia, and who now total more than 15 millions, were attracted thither by commercial opportunities.

(d) *The slave trade* caused the presence of some 60 million Negroes in the Americas today: a shortage of labour for the sugar, cotton and tobacco plantations set up by the Portuguese and Spaniards, and later by the French and British, led to slaves being imported in vast numbers.

(e) *The Hindus* early in history went to the East Indies in search of trade; more recently they have gone to Malaya to work on the rubber plantations and to Natal in the Republic of South Africa.

(f) *The Maltese* have constantly emigrated because of the poverty of the Maltese Islands: many went to Queensland to find work in the sugar-growing industry.

28. Movements in Europe. Movements between countries in Europe have occurred on a much more extensive scale than most people realise. This has happened especially since the Second World War and has stemmed largely from the need for workers following the industrial boom of the past twenty years in the Common Market countries. Figure 7 shows the principal move-

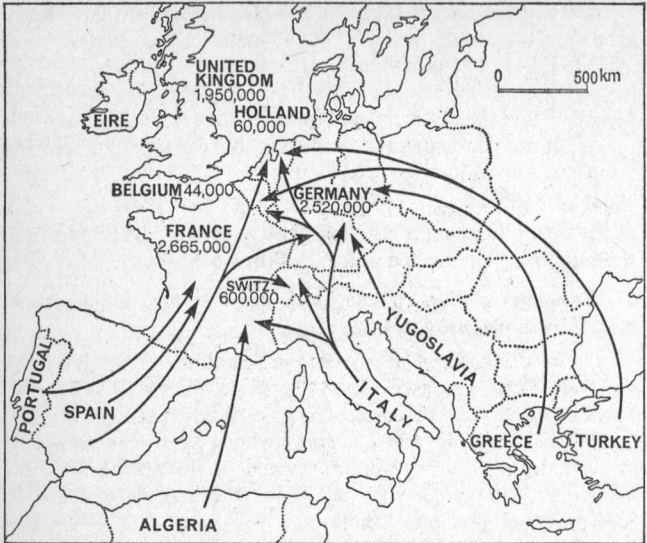

FIG. 7   *Movements of people in Europe.*

The map shows the recent movements of people between countries in Europe, but especially the Common Market countries. The figures indicate the total imported work-force, except for Britain, where they refer to the total number of immigrants.

ments of people in Europe. The figures indicate the total imported work-force in each country since 1945, except in the case of Britain where the figure refers to the total number of immigrants.

(*a*) *France.* The total immigrant workforce is now over 2½ million, consisting mainly of N. Africans (30,652), Portuguese (64,328), Spaniards and Italians. They find work chiefly in building, metal works, agriculture, catering and domestic service.

(*b*) *West Germany.* Some 2½ million foreigners are immigrant workers. They comprise chiefly Turks (599,000), Yugoslavs (514,000), Italians (423,000), Greeks (243,000), and Spaniards (183,000). In addition, West Germany has received many thousands of refugees from East Germany.

(*c*) *Switzerland.* Switzerland, with 600,000 incomers since the Second World War, has the highest proportion of foreign workers

—one in three—of any European country. About 50,000 workers cross the border daily.

(*d*) *The Low Countries*. Belgium and the Netherlands, partly because they are already overcrowded countries, have fewer immigrants although in each case they have admitted around 200,000 and 80,000 respectively. Many Poles work in the Belgian coal mines. The Netherlands has also absorbed considerable numbers of repatriates from Indonesia.

These shifting populations are helping to make a "united states" of Europe a reality.

29. Internal migration. The migration of population takes place on an internal as well as an international scale.

(*a*) *Britain*. Since the First World War there has been a persistent "drift to the south", formerly more especially to London and the Home Counties, but the movement to the south-east still continues and this is causing many problems relating to land use, housing, communications, water supplies, etc. The growing industrialisation of the south-east has been both the cause and the effect of this drift of population. The depopulation of the highland areas, not only in Scotland and Wales but in northern England and the West Country too, has been proceeding steadily for many decades. Some attempts are now being made, as in the Scottish Highlands, to arrest this drift.

(*b*) *The United States*. The "dust bowl" of the 1930s led to thousands of farmers quitting the High Plains region. A large proportion went to California to start anew. Also others who were not farmers were attracted by California's climate and prosperity and, in fact, more than 6 million people have moved into California since 1950. Another movement of tremendous social significance has been the drift of large numbers of Negroes from the south into the urban areas of the north-east. During the decade 1960–70 Florida, Arizona, Nevada and Colorado have witnessed the greatest growth in population with an increase of 25 per cent or more.

(*c*) *The Soviet Union*. At the time of the Revolution (1917) only a very small percentage of the Russian people lived in the Asiatic territories. Even now the proportion is relatively small but during the past fifty years several million people have moved into Soviet Asia. This colonisation, voluntary and involuntary in its nature, has led to the opening up and transformation of many areas. Many large new towns have sprung into existence.

(*d*) *Other countries*. Similar internal movements can be traced in other countries: for example, Norway is experiencing the same depopulation of its highland zone as Scotland, and in France people are leaving the Massif Central and moving into the Rhône Valley and the Paris basin, while in Italy there is a movement from the impoverished south, the Mezzogiorno, to the more developed and prosperous north. No doubt every country, to a greater or lesser degree, is witnessing some internal migration.

While purely social reasons may explain some of these movements, at root they are mostly economic: higher wages, better opportunities, greater variety of work.

THE POPULATION PROBLEM

30. The problem. As was indicated in **16**, the world's population has increased in modern times with such rapidity as to warrant calling this expansion an "explosion". The problem of numbers is already acute in certain areas but perhaps the most serious problem is the threat of greatly enlarged numbers in the none too distant future. Greatly increased numbers will require living space, and food supplies. These aspects we shall deal with in Chapter V. For the moment we can say there are two main considerations with respect to the population problem:

(*a*) The problem of easing local overcrowding.
(*b*) The problem of controlling over-all growth.

31. Solving the problem. Let us take the problem of local overcrowding first. There are three main possible remedies which may be adopted to ease the situation, even though they may not solve it.

(*a*) *Emigration*. In the past large numbers migrated from overcrowded countries, e.g. Italy, but the days of large-scale emigration appear to have gone and few countries will accept immigrants on any scale (the real exception here is between the countries of the British Commonwealth). Where immigration is allowed, it is usually reserved for scientists, technicians and people with specialist knowledge; impoverished, illiterate workmen who have little to offer except their labour are not welcomed.

(*b*) *Development of natural resources*. A more positive approach to the problem is to develop the country's natural resources so that it can support more people. Increased numbers could be supported by such measures as improving agricultural production, fishery development and exploitation of mineral wealth. Such developments, however, usually need capital for their realisation and, all too often, it is the poor countries which are most overcrowded. They are very dependent upon the help, financial and otherwise, which the richer and more developed countries can give them.

(c) *Industrialisation.* The development of manufacturing industry and the production of goods for export offers another possible alternative. Industrialisation is capable of absorbing a certain quantity of surplus rural labour and the export of manufactures provides a source of income. Industrialisation, however, is at best merely a palliative and not a cure for over-population; nor is it always economically feasible or socially desirable. Again, countries lacking industrial raw materials, power resources and technological skill are likely to find industrialisation difficult and of limited applicability.

32. Solving the world problem. From the point of view of overall population growth, there is only one possible practical solution: birth control. This is not an easy matter, for it has to meet the challenge of religious, social and even economic and political factors.

(a) *Religious factors.* Artificial birth control is contrary to the teaching of the Roman Catholic Church, although there has been pressure to try and achieve modification of that Church's attitude.

(b) *Social factors.* Such social customs as ancestor worship among the Chinese, polygamy among the Muslim peoples, and early marriage among the Hindus have, in the past, encouraged population growth, and the continued practice of some of these customs is a factor in current population expansion.

(c) *Economic factors.* The demand for labour after the Industrial Revolution was probably a factor stimulating large families in Britain. Furthermore, the more children there were in a family the greater was the number of wage-earners. This latter factor still applies in underdeveloped countries where children are employed in the fields, tend animals, etc.

(d) *Political factors.* Sometimes national policy has aimed deliberately at stimulating the increase in the population; for example, in the inter-war period, both Fascist Italy and Nazi Germany encouraged the procreation of children by offering state bounties and even medals to prolific mothers!

(*e*) *Health factors.* It has been suggested that hunger stimulates the sexual appetite and, if this is so, it would help to explain the high birth rates in many backward countries. Conversely, a high protein diet, according to one authority, causes low fecundity. Certainly higher standards of living seem to check the natural rate of increase.

It is not likely to be easy to educate people to practise birth control, although there is a gleam of hope in that the use of birth control is beginning to make progress in such overcrowded countries as India. However, it must be remembered that, in any event, birth control must be looked upon as a slow and a long-term method of slowing down population growth.

ECONOMIC ASPECTS OF THE POPULATION PROBLEM

33. The importance of size of population. The size of a country's population is of economic importance for two main reasons:

(*a*) The production of goods and services depends very much upon the amount, availability and quality of human labour, and the supply of this labour is itself closely related to the number, age composition and education of the population.

(*b*) People are consumers as well as producers, and the standard of living bears a close relationship to the numbers sharing the national income, that is, the total of goods and services produced by the economically active population.

NOTE: Also of great importance is the size of a country's population in relation to other factors of production, such as the natural resources, physical handicaps, availability of capital. The total economic output of a country will fall short of what it might be if there is insufficient man-power to make full and effective use of the non-human factors of production. On the other hand, if the population is too large in proportion to the non-human factors of production, then living standards, accordingly, will be much lower than they might be had a better balance between the two groups of factors been obtained.

34. Optimum population. Theoretically there is an optimum population for every country: this optimum is reached when the labour force is just sufficient to make the best possible use of the available resources.

But this fine balance is not necessarily constant; any increase in resources—an improvement in soil fertility, new mineral finds, realisation of power potential or stock of capital—will probably permit or require increased supplies of labour; it follows, therefore, that the level of the optimum population will be raised.

35. Average density. The average density of population of a country is obtained by dividing the total population by the land area.

This gives some interesting results, as the following table shows:

TABLE I.   AVERAGE POPULATION DENSITY (1975 ESTIMATES)

Country	Population (000)	Area (000 km²)	Density (per km²)
Australia	13,502	7,687	1.8
Belgium	9,796	31	316.0
Bermuda	56	0.05	1,120.0
Canada	22,831	9,976	2.5
France	52,913	547	97.0
Iceland	218	103	2.1
India	598,079	3,280	182.0
Rhodesia	6,310	387	16.0
United Kingdom	56,149	245	229.2
Soviet Union	254,382	22,402	11.1
Venezuela	11,993	912	13.0
West Germany	61,832	249	248.0

The average density, which is frequently quoted in textbooks, is often practically meaningless. It fails to indicate the simplest fact that many parts of a country (as in the case of Canada and Australia) are uninhabitable or at least habitable only under extreme difficulty. One of the best illustrations one can give is Japan: the average density for Japan according to the 1974 estimate was 295 per km², but in actual fact only about 16 per cent of Japan's area is habitable, the rest, about 84 per cent, is uninhabitable because of altitude, climate or infertility of soil and so is almost devoid of any economic usefulness; hence the true density of population works out at around 2,000 per km². One must beware, therefore, of accepting the average density at face value.

PROGRESS TEST 2

1. Define the term "race". Name the chief ethnic groups distinguished by the anthropologist and the geographer. **(1, 6)**

2. "There is no such thing as a pure race." Do you agree? **(3)**

3. Which criteria are used to distinguish ethnic groups? **(4)**

4. What is meant by the term "melting-pot of races"? Name an area in the world where large-scale racial intermixture is taking place. **(9)**

5. Explain the meaning of the following terms: demography, Mezzogiorno, *Homo sapiens*, cephalic index. **(2, 5, 10, 29)**

6. Explain what is meant by a population pyramid. How many basic types of population pyramid are there? **(12–13)**

7. What type of pyramid is produced by (*a*) a country with very high birth and death rates, and (*b*) a country with low birth and death rates followed by a recent period of relatively high fertility? **(13)**

8. Which factors helped to keep the population numbers low in the past? **(14)**

9. What is meant by the term "population explosion"? **(15, 16)**

10. Describe the distribution of population in the world, indicating the main regions of high density. **(18)**

11. Which geographical factors have influenced the distribution and density of population on the earth? **(19, 21)**

12. Discuss the relationship between population density and living standards. **(23)**

13. Which motives lie behind the movements of people? **(24–28)**

14. "The shifting populations of Europe are helping to make a united Europe a reality." Do you agree? **(28)**

15. Describe the internal movements of people in (*a*) the British Isles, (*b*) the United States. **(29)**

16. Show how the problems of local overcrowding may be alleviated. **(31)**

17. Why is the size of a country's population important? **(33)**

CHAPTER III

Early Steps in Human Progress

STONE AGE TOOLS

1. The importance of technology. Having looked at the origins, races and distribution of man in the previous chapter, it will be useful next to view the significant stages in his early cultural development. Early man was a food collector and petty hunter and, as such, he could have had no more effect upon his environment than any of the other large predators that were in existence. Early man was very much a part of the natural system. Though at first he lived largely on the sufference of nature, the evolutionary process slowly but surely raised him above other creatures. Man is superior to the other animals because of his intelligence and creative imagination, which enabled him to devise and fashion tools, weapons and mechanical contrivances. When he fashioned his first crude tools, they gave him the beginnings of an ability to influence his environment. But as time went on and his technical expertise grew, he developed the skill to modify his environment to suit his own particular needs, interests and plans. The major innovations in the long process of man's cultural development involved, more importantly, changes in technology for it was these technological advances, above all else, which enabled him to loosen and finally to break the chains of environmental control. As Perpillou has commented: "The search for means to master nature constitutes civilisation in the true sense . . . the standard of civilisation is measured by man's power to control nature" (*Human Geography*).

This chapter is concerned with man's earliest tools, discoveries and inventions—the earliest steps in the long history of his cultural evolution and without which almost everything else that has happened since could not have occurred.

2. The earliest tools. The history of early man's cultural advancement is best appreciated by its relation to the technological stages in his development. In the first of these his tools and weapons were made of stone, wood and bone, especially stone, hence the term the Stone Age. Three principal ages are recognised:

(a) *The Palaeolithic Period*, or Old Stone Age, when man's stone implements were crudely chipped into shape.

(b) *The Mesolithic Period*, or Middle Stone Age, noted in particular for its use of microliths (tiny flakes of flint).

(c) *The Neolithic Period*, or New Stone Age, when implements were finely shaped and ground and polished.

This is the chronological pattern in Europe, but the technological stages were not contemporary everywhere. Also, it will be recalled there are some peoples in some parts of the world that, from the cultural point of view, are still in the Stone Age, incredible as this may seem to sophisticated Westerners.

In the early Palaeolithic times men were collectors and hunters. Associated with this early phase are stones, which bear traces of having been artificially shaped, known as eoliths; these could be grasped in the fist and used as a kind of knuckleduster. Animal bones must have also been used as bone tools are known from middle Palaeolithic times and are associated with what is known as the Mousterian culture. Wood, too, must have been used but since wood is perishable no wooden tools of this early time have come down to us, but it is clear that a branch sharpened to a point at one end would have functioned as a serviceable spear.

Before the introduction of metal, easily chipped stone was essential for the making of tools and weapons. Various kinds of stone were pressed into service but flint, dug out of chalk, was the best material as it readily fractured and gave a sharp cutting edge. Also it enabled early man to fashion a variety of tools, e.g. scrapers, borers, picks and axe-heads. As time went on, a wide range of flint tools were fashioned in ever-improving form and some flints were very beautifully chipped. In Upper Palaeolithic times a number of cultures have been recognised, e.g. Aurigacian, Solutrean, Magdalenian.

During the succeeding Mesolithic Period, when man was still living by a food-gathering economy, several distinct cultures emerged. Hunting and especially fishing seem to have been well-developed as the abundance of microliths and fish-spear barbs attest. The production of microliths, commonly known in Britain as pygmy flints, was associated with the change in ecological conditions at the end of the Ice Age, i.e. there was much melt-water around which promoted fishing.

3. The Neolithic Period. The New Stone Age followed the Mesolithic Period and lasted until the introduction of metal working. The Neolithic Period introduced a new feature into the working of stone, the process of grinding and polishing. The highly-finished stone implements which characterise the later phase of the Stone Age contrast with the ruder workmanship of the Palaeolithic Period. During the Neolithic Period plant and animal domestication began and the first agriculture was developed. The outstanding feature of the Neolithic Period is, in fact, the emergence of a new type of economy which was wholly or partly based upon domesticated crops or animals, or both. This change in the economic system has been styled the Neolithic Revolution, although in some parts of the world the change was far from being sudden and took a long time to effect. The Neolithic Period is also associated with the development of various crafts among the more important of which were pottery, basketry, spinning and weaving. Although permanent settlements associated with fishing communities had appeared in Mesolithic times, settlements of any size and with a developed social organisation belong to the Neolithic Period.

THE WORKING OF METALS

4. The discovery of metals. Metals are commonly distinctive materials possessing distinctive physical properties such as hardness, heaviness, lustre, etc. Comparatively few metals, however, occur in a "native" or "free" state (gold, silver and copper are the most common); more usually they exist as compounds, that is, they are chemically combined with some other mineral as, for example, oxygen, sulphur, silicon. Seldom do such compounds resemble metals. This helps to explain why man was so long in discovering metals. Man, moreover, could not progress from the Stone Age into the Metal Age until he had learned how to extract metals from their compounds. His discovery of how this could be done marks one of the most important steps in his cultural development. There seems little doubt that gold was the first metal known for its bright yellow lustre and its occurrence as nuggets or dust in stream beds would easily attract man's attention. Copper, too, often occurs in a natural state and it is usually tarnished a little—rubbing or scratching will reveal its warm, ruddy colour—and it also became known very early on.

5. The Chalcolithic Period. This term is commonly used for the Copper Age, the period when copper-working was known, but during which, stone was still used for most tools. Although it is generally believed that gold was the first metal known to man and was used for personal adornment, the earliest metal from which articles of use were fashioned was copper. Copper exists in the native or natural state, and in earlier times must have been found on the land surface in greater quantities than it is today. Copper was known *c.* 5000 B.C. since copper beads were found in Badarian (a pre-dynastic culture in Ancient Egypt) graves, but copper objects did not become common until Middle Pre-dynastic times in Egypt, i.e. the Gerzean Period. Copper, being malleable, is easily hammered into shapes and it must have been used in this way long before man learned how to melt it and cast it into moulds. Just where and when the art of casting copper arose is not known precisely but it seems to have been known in Egypt by the Gerzean Period since quite elaborate tools have been found in graves of that time. Once it was discovered that copper could be melted and be run into moulds, the art of metallurgy was born, another epoch-making development in man's cultural progress for it ushered in the Metal Age in which we are still living.

6. The Bronze Age. Bronze is an alloy of copper and tin; the amount of tin present varies up to about 12 per cent. The addition of tin to copper produced a material which was rather harder than copper and so was more suitable for making tools and weapons. Just how, when and where man discovered that the addition of tin to copper would produce this useful alloy is a matter of much uncertainty, but it is generally thought that the discovery was accidental, especially since copper and tin ores are seldom found in common association. The archaeologist Woolley discovered some bronze objects at Ur but it seems likley that the people of Ur were not conscious that they had made these objects from an alloy. These objects date from the fourth millenium B.C. Bronze was not generally known until around 2500 B.C., although by this time or thereabouts it was in use in Mesopotamia. Bronze seems not to have been introduced into Europe until about 2000 B.C. or perhaps a little later. The Iberian peninsula and Ireland became important centres of bronze production. In Britain the Bronze Age is commonly believed to have existed from about 1900 to 500 B.C.

7. The Iron Age. The Hittites of Asia Minor are commonly credited as being the first iron-workers; certainly they possessed much knowledge of metals. They appear to have been using iron before 1500 B.C. but they guarded their secret of iron-working until their empire suddenly collapsed c. 1300 B.C. after which time the knowledge of iron gradually spread across Europe. It was known in ancient Greece by the eleventh century B.C. and had reached central Europe by the eighth century B.C. The use of iron led not only to improved tools and weapons but also to a general cultural advance. During the early Hallstatt period of the Iron Age (c. 700–600 B.C.) Celtic-speaking communities spread over much of central Europe and were responsible for the intensification of trade. As a trading network developed, Iron Age culture was diffused over much of the continent so that not only were there similarities in language and artistic styles, but also in hill-fort defences. The later Iron Age period, known as La Tène, saw the introduction of wheel-made pottery and gold coinage.

THE FIRST AGRICULTURAL REVOLUTION

8. The development of agriculture. If tool-making was the first major step in human progress, the development of agriculture was the second major innovation. There are signs that it existed c. 7000 B.C. in the Near Eastern area. It grew out of the collecting and hunting economy which preceded it, a primitive economy which continues to prevail still in a few parts of the world. The invention of agriculture had immense human geographical repercussions because it led to the development of a sedentary way of life which stands in marked contrast to a nomadic existence. Agriculture permitted the production of food surpluses which could be stored against lean times and which could sustain man until the next harvest. Prior to this, man had been compelled to wander around in an unrelenting search for food. Since the growing of crops implied a settled existence, other developments issued from community life: an assured food supply, the division of labour, permanent dwellings, the collection of personal possessions, the beginnings of trade, and the growth of leisure. From these developments the first civilisations emerged.

9. The Middle East as the cradle of agriculture. The traditional view was that the Middle Eastern area was the home of agriculture since it was here that archaeologists discovered the oldest civilisations which had practised crop growing, and here that botanists found wild wheat and barley growing from which the cultivated varieties had been developed. Support was subsequently lent to this traditional view by the Russian scientist Vavilov who argued that seed agriculture—the cultivation of wheat, barley, millets and sorghums—probably originated in hill lands at various places in the Middle East and that this knowledge was spread to adjacent alluvial lowlands. It is now reasonably well established that seed planting (and probably herd animal domestication too) had its origins in the Middle Eastern area. Certainly food production, especially the growing of wheat and barley, was present in Egypt 7,000 years ago and in some areas, as recent archaeological evidence suggests, was practised even earlier.

10. Carl Sauer's theory. Different theories on the origin and spread of agriculture have been advanced, but most of them are not strongly supported by evidence. One of the most attractive and original is that put forward by a distinguished American geographer, the late Carl Sauer. He averred that, contrary to traditional and popular belief, the first agricultural innovations were in all probability root planting and the planting of cuttings rather than seed planting, and that the area in which this occurred was south-east Asia (*Agricultural Origins and Dispersals*). According to Sauer, seed planting and the domestication of herd animals were later developments, although he does accept that cereal cultivation had its origin in the Middle East, whence it subsequently spread into India, Europe and north Africa.

Sauer questions some of the assumptions upon which a Middle Eastern origin of agriculture is based:

(*a*) That sedentary life was contingent upon the change from food gathering to food production; he argues there must have been settled life *before* the change could occur.

(*b*) That a growing shortage of food and hunger was an incentive to the change; he suggests, on the contrary, that food production was likely to arise in regions of plenty where there was a surplus of food.

(*c*) That agriculture began in semi-arid regions; he suggests it was more likely to occur in tropical forest lands and he thinks south-east Asia was the cradle of agriculture.

Sauer selected south-east Asia as the probable originating area since here would be found Mesolithic fishers living in sedentary communities. Here, a simple form of agriculture could have emerged, he argues, based on the planting of tubers with a digging stick. Tubers could be broken off edible roots and put into holes in the soil dug with a crude digging stick, a practice still followed by the simplest cultivators of the tropical forest lands. Sauer claims that this kind of cultivation is simpler and probably much older than the growing of grain by scattering seeds on ploughed soil. Cereal cultivation to Sauer's way of thinking is a relatively advanced economy and was a later development.

11. American agriculture. Before the discovery of the Americas by Europeans, Indian cultures seem to have evolved independently in isolation from other peoples. Agriculture seems to have been a native development since the plants which were cultivated were species native to the New World. "The invention of agriculture in the Americas," writes Philbrick, "is also divided into two phases by the same kind of agricultural innovations which took place in the Old World—root planting and seed planting" (*This Human World*). Seed planting probably had its origin in southern Mexico where maize, beans and squashes were grown, while root planting (potatoes, manioc) occurred in the northwest of South America and in the Andean Highlands.

THE DOMESTICATION OF ANIMALS

12. Domesticated animals. The domestication of animals was an innovation of the Neolithic food producers and it appears to have been contemporaneous with, or a little later than, the introduction of plant cultivation; at any rate animal domestication was closely associated with cultivation. It is very probable that in the course of human history experiments in domestication were carried out on large numbers of animals but the number of species successfully domesticated remains small—less than forty; nevertheless, many are found in the earliest farming communities in the Middle East. As Jones says: "the Middle East is certainly a focal point in the wide region where wild species of domesticated animals occur and within which domestication could have originated" (*Human Geography*).

The dog, cow, sheep, goat, pig and ass were the first animals to be domesticated. With the exception of the dog, and the ass these were all food animals. Horses, camels and yaks were domesticated later and were primarily beasts of burden. Several birds, e.g. chicken, duck, goose, peacock and pigeon, were domesticated and provided food in the form of eggs and flesh, while two insects, the bee and the silkworm, were brought into the service of man. All these were creatures of the Old World. In the New World there were fewer animals amenable to domestication, but the llama and alpaca were domesticated in the Andean region and were used as pack animals, while the turkey came from Mexico. Since these early domestications, man has produced a great variety of breeds within many of the species, but has hardly added a single other creature to the list of domesticated animals.

13. Methods of domestication. "The conditions necessary for domestication include: gregariousness for herd animals, docility, a liking for man, love of comfort and the ability to breed in captivity. Few animals fulfil all the conditions" (Jones, *Human Geography*). It is difficult to know with any sureness precisely how the process of domestication came about. Some animals may have been caught and tamed and subsequently domesticated, much as used to happen in India where wild elephants were rounded up and tamed and then trained to work. Some authorities would go along with the archaeologist Childe who believed that the growing desiccation in pre-historic times compelled men and animals to congregate around shrinking water-holes and "this enforced juxtaposition produced the symbiosis we know as domestication" (*What Happened in History*). Yet, others believe the early stages of domestication may have been linked with religious ritual; for example, cattle were connected with ritual in the early civilisations of Egypt and Mesopotamia. But the whole problem is rather vexed.

14. The use of animals. The first animal to be domesticated was the dog and it was first used to help man in the chase. Many believe that the dog had become man's companion as early as Upper Palaeolithic times; certainly it had become domesticated by Mesolithic times since bones gnawed by dogs have been found in the kitchen middens of Mesolithic times in Denmark. Cattle, sheep and goats seem to have been brought under the control of man at an early date, probably before 5000 B.C. and perhaps even between 7000 and 6500 B.C. They provided flesh, milk, hides, wool and hair. A stone frieze, unearthed from the site of Ur in Sumeria, depicts cattle being milked. The first beast of burden was the ass for it was being used in ancient Egypt and Sumeria before 3400 B.C. and could have been domesticated at a considerably earlier date. The horse was probably first tamed in the steppes of Central Asia but there is no mention of it until about 2000 B.C. The horse was, however, known to the Hittites who arrived in Asia Minor just after 2000 B.C. The camel had been introduced into Egypt roughly around the same date where it came to be used as a beast of burden, but it seems likely it was first domesticated in the Gobi region of eastern Asia about a millenium earlier.

THE EARLY APPARATUS OF CIVILISATION

15. The effects of the agricultural revolution. The introduction of agriculture effected a revolution in the life of man. As already mentioned, crop cultivation assured him of a sufficient food supply, which also allowed him to increase his numbers, and permitted him to live a settled existence in small communities. This fundamental revolution also necessitated great changes in his tools and in his life-style. Tilling the soil, harvesting the crops, removing the grain from the chaff, and storing the grain demanded new implements and utensils. Thus, the introduction of agriculture brought with it the introduction of farm implements, of a rudimentary apparatus for grinding corn, of pottery, and of houses. It also demanded new attitudes of mind: the necessity to work long and hard, especially at certain times during the year, the need to develop thrift so that he would have a supply of seed-corn for the next planting, the need to organise and plan his work routine.

16. Agricultural implements. Man's first tools had been designed for food-gathering and hunting, not agriculture, and one might have expected that the introduction of such a novel activity as crop cultivation would have led to the invention of a new set of implements, designed especially for its performance. But this was not so; rather, as has usually been the case, man has adapted an old implement to a new use; hence, his earliest farm tools were adaptations of the tools with which he was familiar.

Perhaps the earliest, and certainly the simplest, of the implements man used in his agricultural activities was the digging-stick, a tool still used by many backward people. The digging-stick was merely a pointed stake, the end of which was sometimes hardened by fire. The digging-stick was used to gouge a hole in the soil or to scratch its surface. Clearly it was a primitive and not very effective instrument, but in due course it developed into a spade. An interesting intermediate type was formerly used by the Maoris of New Zealand: this consisted of a pointed stick to which, near the sharpened end, a transverse piece of wood was fixed to serve as a foot-piece.

The hoe was the tool most generally used in early cultivation. In its simplest form the hoe has a sharp point and is more like a pick but in its developed form the pick point has expanded into a blade similar to an adze. In many parts of Africa the hoe is still widely used for tilling the soil. The hoe was used in ancient Sumeria and Egypt and it continued to be used as the chief agricultural implement in northern and western Europe until the beginning of the Christian era.

In Egypt the hoe gave way to the plough at an early date for there are wall carvings of primitive ploughs c. 2750 B.C.

17. Pottery. "The preparation and storage of cereal foods," wrote Childe, "may be supposed to have put a premium upon vessels which would at once stand heat and hold liquids. A universal feature of Neolithic communities seems to have been manufacture of pots" (*Man Makes Himself*). Pottery and fragments of broken pottery occur in most Neolithic sites and attest to potmaking on a large scale. Yet pot-making is a fairly complex process and involves the use by man of a chemical change. In the earliest stages of pottery a vessel was built up of plastic clay by hand and fired on an open hearth. Later the potter's wheel and the kiln were introduced, followed by the use of glazing. The earliest pots were, it seems likely, imitations of containers made from other materials such as gourds, bladders, basketry, etc. Building up a pot, however, was a supremely creative act and once man had become familiar with the properties of his raw material he could experiment and design pots of varied sizes and shapes. Pots were used for storing grain, dried fruit, water and wine.

18. Textiles. The making of cloth through the arts of spinning and weaving is a very old craft and archaeologists have discovered evidence of a textile industry in the remains of the earliest villages in Egypt and the Near East. Linen cloth made from flax fibre seems to have been the first to be introduced. By 3000 B.C. wool was also being used in Egypt and Mesopotamia and it is known that cotton was cultivated in the Indus valley soon after that date. Textile products and the wooden mechanisms used in their manufacture can only be preserved under very exceptional conditions but wall paintings and carvings of later times demonstrate that the arts of spinning and weaving had been developed. Some sort of loom must have been devised in order to produce cloth and Childe wrote "in the Old World a true loom goes back to neolithic times . . . The invention of the loom was one of the great triumphs of human ingenuity" (*Man Makes Himself*).

PROGRESS TEST 3

1. How far would you agree that technological progress has been the mainspring of human progress? **(1, 3, 4, 15)**

2. Name the three main phases of the Stone Age and indicate the chief characteristics of the stone tools used in these periods. **(2, 3)**

3. "The discovery of metals marks one of the most important steps in man's cultural progress." Discuss. **(4–7)**

4. What changes in man's way of life issued from the invention of agriculture? **(8)**

5. Outline Carl Sauer's theory on the origin of agriculture. Do you think his argument is convincing? **(10, 11)**

6. Elaborate upon the following statement: "The introduction of agriculture effected a revolution in the life of man." **(15)**

7. Briefly explain the meaning of the following terms: microlith, eolith, digging-stick. **(2, 16)**

CHAPTER IV

Social Geography

HUMAN CULTURE

1. What is culture? The term "culture", like the term "society", is very difficult to define. The two are intimately associated, for every society possesses a particular, and often a very distinctive, culture. The term "society" implies two things:

(a) An organised group of human beings.

(b) A group with a distinct culture.

A society is probably best defined in terms of its culture; hence it is desirable that we should be clear as to the meaning of culture. Broek and Webb have simply but effectively defined a specific culture as "the total way of life of a people". To the anthropologist culture often means a stage in civilisation. The assemblage of ideas, beliefs, institutions, skills, tools and artifacts possessed by a people at any stage in time constitutes its culture. "The content of each culture includes," write Broek and Webb, "systems of belief (ideology), social institutions (organisation), industrial skills and tools (technology), and material possessions (resources)." (*A Geography of Mankind.*)

2. Culture—a dynamic. Although culture might be differently defined, the foregoing description adequately suits our present purpose. It must be stressed, however, that culture is seldom, if ever, stagnant: it is usually in a constant state of change. The Australian aborigines, when they were discovered, were in the Stone Age and had been in this condition for several thousands of years. Clearly, the changes in their culture had been very slow indeed—in fact, had scarcely altered throughout several millennia—but we can never say that their culture had remained static, had not changed. In this particular case the cultural changes were extremely slow. At the present day, however, cultures the world over are in a constant state of change—they are dynamic.

3. The Eskimos: an example of cultural change. To illustrate the fact that cultures are in constant process of change, let us look for a moment at the Eskimos. Less than a century ago, and in some instances less than fifty years ago, the Eskimos, who inhabit the arctic fringes of North America, lived isolated, self-contained lives expertly using the restricted resources of their homeland but living a precarious existence. Until recent times, their food, clothing, shelter, tools, equipment and boats, everything needed for existence in fact, had to be obtained or made locally. Since they came into close contact with the white man, however, their lives have been very much changed. Often, the native costume has been modified, new foods have been introduced, and new weapons, such as the gun, have been adopted. Nowadays, because the Eskimos can get paraffin for their lamps and their stoves, they throw away the blubber which once was their most prized resource. Many speak English, many chew gum and quite a number have forsaken their fishing and hunting and work for the Americans and Canadians as miners, technicians, radio-operators, etc. Within two or three generations their lives and culture, in many cases, have been radically transformed.

4. Factors of cultural change. This brief account of the Eskimos will serve to show how cultural change occurs. Formerly the Eskimos had to live by their wits, their skills and the things they made for themselves; they had to design and invent clothes, dwellings, weapons, means of transport, etc., which suited the environment in which they lived and they had to make use of a very limited supply of natural resources.

When the Eskimo came into contact with white men he copied, borrowed, stole and bartered. He learned white men's ways, even adopted some of their bad habits, such as drinking spirits, and came to possess new equipment and tools, e.g. stoves, guns, which made life easier.

Thus cultural change may develop in two ways:

(a) By discoveries and inventions within a society.
(b) By the introduction of something new from other societies.

NOTE: It does not necessarily follow that a new introduction will be accepted and adopted by a society. An innovation must be accepted before it can become part of the group's culture.

5. Evolution versus diffusion. Whether man progresses culturally through his own efforts and independent invention or whether he advances as a result of cultural contacts with, and borrowings from others, has long been debated. There are some who believe in man's inherent creative ability who, when faced with a problem, will find ways and means of overcoming it; contrariwise, there are others who believe that man's aptitude for discovery and invention is limited and that cultural anthropology can show us numerous cases of man never rising to the occasion even when the pressures were great. Let us look briefly at the arguments for each.

6. Evolution. Man, unlike the animal, is a rational creature: he has the power to think and to reason. Thus, if he is confronted with a problem, he has the power to tackle it and perhaps to solve it: clearly, he will not solve all his problems, he will not be successful every time. "Necessity is the mother of invention,' it is often said: this is by no means true, although necessity on many occasions must have led to invention. "One who believes that the natural environment conditions, if not determines, the way of life, probably will emphasise local invention as man's response to nature's challenge" (Broek & Webb, op. cit.). Let us list some of the arguments of the evolutionists.

(*a*) Fire is one of the greatest, most basic and earliest discoveries of man. To believe that fire was created only once and that the idea then spread around the world is scarcely believable and tenable. Fire may be started in many ways (e.g. by the sun, lightning, volcanic action, etc.) and man must have seen and learned to use fire many times in many places.

(*b*) Both the ancient civilisations of the Old World and the New World invented calendrical systems but they are different from each other, clearly suggesting quite independent invention.

(*c*) While all ships in the West evolved from the dugout canoe, the Chinese junk is of a completely different construction, having been derived from the raft. Here are two quite distinct and original forms of nautical invention.

(*d*) There are known cases of simultaneous independent discovery and invention: Priestley and Scheele discovered oxygen independently and Marconi and Popov invented radio independently.

(*e*) The American Indians of pre-Columbian times discovered the art of agriculture independently and they cultivated crops unknown in the Old World. They also domesticated the llama, alpaca, turkey and guinea pig, animals unknown in the Old World.

7. Diffusion. The spread of cultural traits from one place to another is termed diffusion: this may result from the direct contact of peoples or it may happen indirectly through a series of intermediary transmitters. The diffusionist theory gained great support through the writings of G. Elliot Smith and W. J. Perry some half a century ago. Perry, however, carried things to an extreme and claimed that all early cultural elements emanated from Ancient Egypt. Many cultural traits seem to support the diffusionist theory:

(*a*) *The art of cereal crop cultivation* can be traced from an early centre in south-west Asia to other parts of the Old World.

(*b*) *The magnetic compass* which suddenly appeared in medieval Europe in all likelihood was an importation from China where it had long been known.

(*c*) *The practice of smoking*, which is now almost universal, was introduced into Europe from America and then spread to other parts of the earth.

(*d*) *The oculus or "eye"* on the prows of ships seems to have been an ancient Egyptian introduction but has a wide occurrence amongst primitive craft throughout the world.

(*e*) *The horse* was introduced into the Americas by the Spaniards and the Great Plains Indians adopted the horse and the art of riding.

8. Conclusion. Many more examples and arguments could be adduced to support both theories but enough has probably been said to illustrate them. It is difficult, if not impossible, to say which of the two is correct. The truth is probably neither: in all likelihood both have occurred. It must be admitted, however, that, generally, introduced traits, rather than local inventions, are responsible for most cultural changes.

CULTURE AND GEOGRAPHY

9. Man versus nature. Natural factors, such as geographical position, structure, relief, soil, climate, water supplies, vegetation and animal life influence, and sometimes limit and control, human activities. While we no longer believe in determinism, i.e. that man is fundamentally controlled and conditioned by his environment so that his way of life is strictly limited and determined for him, we cannot ignore completely the effects which the natural environment has upon man. The philosophical doctrine of possibilism, i.e. that man is largely the master of his own fate since the environment offers him a choice of possibilities of which he may or may not take advantgage, was espoused by the French school of geographers, especially Vidal de la Blache and Lucien Febvre. This concept has now taken over from determinism, which has few exponents at the present day, but it is equally necessary to be wary and not to push possibilism to an extreme for there are still many environmental influences which restrict man's activities. Some modern geographers, such as the late Sir Dudley Stamp, have stressed this and Professor O. H. K. Spate has postulated the idea of probabilism which suggests that though there are possibilities everywhere, some are more probable than others—a definite implication that in certain circumstances nature still calls the tune.

10. Man's growing ascendancy over nature. Man can develop and turn many of the natural conditions to his own advantage. Though unable to alter materially many of the physical conditions of his natural environment, he has at least been able to mitigate and modify some of their effects, such as the following:

(a) Physical obstacles have been surmounted by building bridges and tunnels.

(b) Steeply sloping land has been made cultivable by terracing.

(c) Marsh, fen and floodland have been reclaimed by drainage.

(d) Poor and useless soils have been turned to good use by soil mixing and fertilisation.

(e) The problems caused by lack of rainfall or light rainfall have been solved by the use of irrigation.

(f) Certain climatic problems have been overcome by the use of glasshouses, central heating, air-conditioning and refrigeration.

(g) Agricultural output has been increased by plant and animal breeding.

(h) Insect pests and the like are being largely eliminated by the use of pesticides, etc.

(i) Time and distance have been drastically modified by radio and telecommunications and new forms of transport.

These, and many more, activities of man have done much to lessen natural controls and influences. As man's scientific knowledge and technical developments increase, the bonds of nature will be increasingly lessened. Just how far he will be able to overcome nature only the future can tell.

11. The cultural landscape. Sometimes a distinction is drawn between the landscape prior to human occupation, called the natural or physical landscape, and that which has resulted from human settlement and man's activities, called the cultural landscape. (The cultural landscape or humanised landscape, as it is sometimes alternatively termed, bears the imprint of man's occupation and utilisation of it. All man-made features, such as buildings, hedges, plantations, roads, pylons, reservoirs, quarries, etc., are called cultural features.)

Sometimes it is difficult for us to appreciate the extent to which man has modified the natural landscape. Because so many of us have been brought up in towns, we tend to think that the rural areas of the countryside are natural landscapes. We forget that the pattern of fields, the trim hedges, the clump of trees, the picturesque cottages have all been made by man and that very often the countryside is just as "artificial" as the town in which, perhaps, we live.

Increasingly in Britain, and, indeed, in most areas of the world, human interference with the natural landscape is going on apace, and man's imprint upon the physical environment is becoming stronger, and he is in many cases destroying the beauty of the landscape and making it ugly.

12. Cultural features. The chief man-made, or cultural, features may be grouped into the following categories:

(*a*) *Buildings:* e.g. dwellings, office blocks, factories, warehouses, theatres, educational institutions, hospitals, churches and monuments.

(*b*) *Features of farming:* e.g. fields, field-walls, hedges, orchards, plantations, dew-ponds, cattle-grids, wells and windpumps.

(*c*) *Features connected with extractive industries:* e.g. quarries, adits, spoil-heaps, mineral workings and dredges.

(*d*) *Drainage features:* e.g. ditches, drainage channels, straightened rivers and levees.

(*e*) *Features connected with water supply and conservation:* e.g. reservoirs, water-towers, pumps, irrigation channels and aqueducts.

(*f*) *Communications:* e.g. roads, railways, cuttings, embankments, bridges, tunnels, canals, airfields, telegraph poles and wires, radio masts and pipelines.

(*g*) *Features of marine engineering:* e.g. groynes, breakwaters, piers, docks, locks and oil-rigs.

(*h*) *Recreational features:* e.g. parks, football and cricket grounds, golf courses, bathing pools and chair-lifts.

THE LANGUAGES OF MANKIND

13. The origin and importance of language. The development of speech and language is one of the fundamental developments of man's cultural evolution. Language is a facet of human culture like social organisation or religion: it is a cultural trait. It is without doubt the most important item of social equipment for living for two reasons:

(*a*) Only by speech and gesture can man communicate with his fellow men.

(*b*) It is the chief means of transmitting man's cultural heritage from one generation to another.

The communication of thought and ideas has been of tremendous significance since the very earliest days of human existence; today it is more important than ever before. Not even the simplest human societies can function unless men can communicate with one another. Imagine the limitations placed upon the individual (and upon social groups) if there were no such thing as language!

14. A Babel of tongues. Because primitive man wished to express his thoughts and feelings he used his vocal chords to produce a series of grunts and squeals which, in due course, developed into speech or language. Just how systematised language evolved is difficult to say but we are all familiar with the way in which new words and expressions are introduced and become accepted, e.g. "wireless", "oil slick", "stereo".

Over several millennia thousands of languages have developed. The study of their linguistic characteristics, their relationships, their distribution and their significance from the cultural point of view is fascinating but complex.

Many languages bear obvious traceable relationships and, accordingly, would appear to have been derived from a common root tongue. Such languages are said to belong to a linguistic family; for example, most European languages belong to the Indo-European family. A dialect is merely a fairly local variant of a particular language differing from it in certain aspects of pronunciation, vocabulary and grammar. Some languages have died out or may no longer be used for general purposes, e.g. Latin, Cornish.

15. Classification of languages. Geographers are interested in the genetic classification of languages since this indicates relationships by origin and development. Some sixteen principal linguistic families or groups may be distinguished. Some of these groups may include sub-groups and within each sub-group there may be several distinct languages; for example, the Germanic sub-group of the Indo-European family includes German, English, Dutch, Flemish, Frisian, Danish, Swedish, Norwegian, Faroese and Icelandic. Figure 8 shows the broad distribution of the main linguistic family groups in the world. Table II gives a simplified classification of European languages.

16. The practical importance of languages. Language is of great importance from many points of view:

(*a*) *A common language acts as a potent cementing factor in a social group.* It welds the peoples into a community and engenders mutual understanding; hence its importance in creating nationality (*see* IX). Conversely, where differences of language prevail, whether between or within states, an obstacle to understanding, co-operation and unity is set up.

(*b*) *Business, clearly, makes great demands upon communication, but language is especially important.* Simple barter can be carried on without the use of language but large-scale, organised trading would be impossible without the medium of language, both spoken and written. The Phoenicians are credited with the invention of the alphabetic system of writing simply because their commercial transactions and accounting needed a written language.

(*c*) *The study of linguistic affiliations is able to throw light on human movements and contacts in the past.* For example, it is claimed that a North American Indian tribe, the Mandans, understood Welsh and had some Welsh words in their speech suggesting a contact with Welsh folk long before Columbus discovered America.

17. The languages of commerce. Unfortunately, man has developed a multiplicity of tongues and these are a great hindrance to commercial intercourse between peoples. Largely for this reason, a few languages, mostly European languages, have come to be used as "the languages of commerce" and most international transactions are carried out in these languages.

Man prefers to do business in his own language or in one with

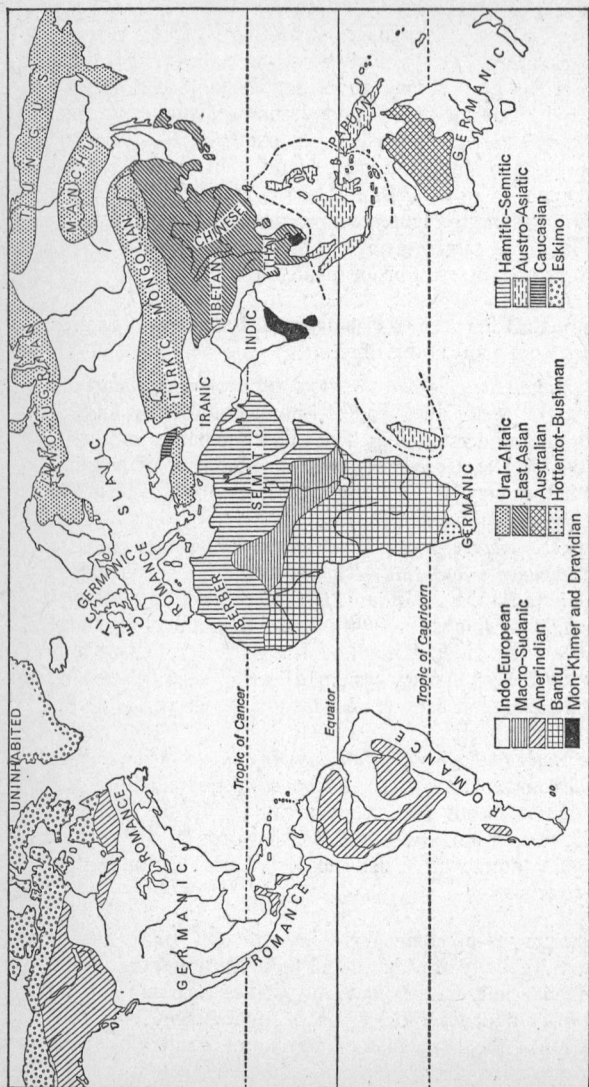

FIG. 8 *Linguistic families of the world.*

The world is truly a Babel of tongues. The map shows only the main linguistic families: each family may contain many distinct languages. For example, in the Indian sub-continent alone there are several dozen languages—this is partly why English was for so long the official language in the area. Of all the languages in the world English is the one most widely spoken.

TABLE II.   EUROPEAN LANGUAGES

1. *Indo-European*
   (*a*) Greek

   (*b*) Romance
      (*i*) Italian
      (*ii*) French
      (*iii*) Spanish
      (*iv*) Catalan
      (*v*) Portuguese
      (*vi*) Romanian

   (*c*) Germanic
      (*i*) German
      (*ii*) Frisian
      (*iii*) Flemish and
         Dutch
      (*iv*) Danish
      (*v*) Swedish
      (*vi*) Norwegian
      (*vii*) Icelandic
      (*viii*) English
      (*ix*) Faroese

   (*d*) Slavonic
      (*i*) Polish
      (*ii*) Czech
      (*iii*) Slovak
      (*iv*) Slovene
      (*v*) Serbo-Croat
      (*vi*) Macedonian and
         Bulgarian
      (*vii*) Russian
      (*viii*) Ukrainian
      (*ix*) White Russian
      (*x*) Ruthenian

   (*e*) Baltic
      (*i*) Latvian (Lettish)
      (*ii*) Lithuanian

   (*f*) Celtic
      (*i*) Gaelic
      (*ii*) Welsh
      (*iii*) Erse
      (*iv*) Breton

2. *Ural-Altaic*
   (*a*) Uralic
      (*i*) Lapp
      (*ii*) Finnish
      (*iii*) Estonian (Esth)
      (*iv*) Magyar (Hungarian)

   (*b*) Altaic
      (*i*) Tatar
      (*ii*) Turkish

3. *Albanian*

4. *Basque*

5. *Maltese*

NOTE: There is some dispute as to the true affiliation of both Albanian and Basque.

which he is familiar. It is very largely because European peoples spread throughout the world, conquering or colonising various parts that the European languages have become predominant in international usage.

The fact, too, that many of the new developments in communications were invented in Europe has tended to emphasise the importance of the European languages; increasingly, with the greater use of telegraphy, telephones, etc., the languages of commerce have tended to become restricted to a few of the European tongues.

The chief commercial languages are the following:

(a) *English* is the most widespread of all; not only is it the mothertongue of the British peoples in the United Kingdom, Australia, New Zealand and Canada, but it is the language spoken in the United States, in South Africa and in many areas which were formerly part of the British Empire.

(b) *French*, which formerly was the language of culture and diplomacy in Europe, and which was widely understood in North and Central Africa and south-east Asia, has lost ground as France has declined as a great power and as it has lost its colonial empire. We should note the presence of a large French-speaking minority in Canada.

(c) *Iberian languages* have currency throughout practically the whole of Latin America. Spanish and Portuguese are, of course, European derived languages. With the major exception of Brazil, where Portuguese is the official language, Spanish predominates generally throughout Central and South America.

(d) *Russian* is growing in importance but is largely confined to the communists countries. Russian has been systematically spread not only through the European territories of the Soviet Union but throughout the Asiatic territories where among the non-Russian peoples it is being compulsorily taught as the second language.

(e) *German*, though widely used in Central Europe and regarded as the "scientist's language", has not developed into an important commercial language except in Europe.

18. Lingua franca. "The language of commerce, when carried on between peoples speaking different tongues, is generally of a very mongrel character. In the days when Italian trade was predominant in the Levant, there arose in all the coasts of that region a trade language, the basis of which was a corrupt Italian, but which borrowed numerous words from the local dialects in different places. This language is known as the lingua franca, and is still spoken in many of the eastern Mediterranean towns. The dominant languages of commerce at the present day have all begotten corrupt forms of speech of a similar nature." (*Chisholm's Handbook of Commercial Geography*). The following are some examples:

(*a*) *Pidgin English* is a jargon of English, Portuguese and Chinese words arranged in accordance with Chinese syntax; it is widely spoken in eastern and south-eastern Asia.

(*b*) *Swahili* is a language derived from Bantu mixed with Arabic, English, Portuguese and Hindustani. It is widely spoken throughout Kenya and Tanzania and even among some of the tribes of the Zaïre basin.

(*c*) *Creole French*, of which there are many varieties, occurs in Louisiana and Haiti. In Haiti, the original pidgin has been adopted as the language of the country.

THE RELIGIONS OF MANKIND

19. Definition of religion. "Religion" is well known for being difficult to define and it is doubtful if any completely satisfactory definition has yet been given. With the exception of simple trival religions, most religions may be said to attempt to explain the mystery of life and involve on the part of man an attitude of reverence to a Supreme Being together with an associated code or system of behaviour which embraces worship in one form or another.

In all likelihood religion had its origins in nature worship. To primitive man, with his limited powers of understanding and reasoning, the earth appeared to contain many things possessing mysterious powers, e.g. thunder and lightning, volcanic eruptions, ebbing and flowing wells, and these he believed to be the manifestations of spirits. Accordingly, he came to think of his world as being full of spirits or deities which demanded propitiation by worship and sacrifice.

Out of this arose polytheism (belief in many gods), particular peoples choosing particular deities which they regarded as being especially favourable to them, e.g. coast dwellers were likely to single out for special honour the god of the sea. In, for example, the Brahmanism of the Hindus, the various deities came to be looked upon as differing manifestations of one God. Then Judaism proclaimed the belief of a single omnipotent God, and Christianity and Islam followed Judaism in teaching a faith in which monotheism (belief in one god) was absolutely fundamental.

20. The principal religions. Excluding tribal religions, there are about half a dozen distinctive religions in the world—principally Christianity, Islam, Judaism, Buddhism, Hinduism and Shinto, and Confucianism and Taoism in the Far East. Although most religions have particular areas of occurrence, religion is not bound by any political frontier or geographical barrier. Christianity, for instance, has its adherents in almost every corner of the world. Let us look, briefly, in **21–26** below, at the origins and adherents of the main religions of mankind.

21. Christianity. Founded upon the teachings of Jesus of Nazareth in the first century, Christianity claims the largest number of adherents of any of the world religions, something of the order of 1,000 million. Like most religions it has split up into sects, and there are three main branches of Christianity.

(a) *The Roman Catholic Church*, the original Christian Church founded upon the teachings of Christ and propagated by his disciples, notably St. Peter. The Pope or Holy Father in Rome became the acknowledged head of the Western Church. The Roman Church has always had a pronounced missionary zeal and made converts in the Americas, Africa and the East. Today there are over 600 million Roman Catholics. Outside Europe, the largest following is in North and South America. The Roman Church has always had a strong hold upon its adherents and in Latin America Catholicism still has a powerful influence.

(b) *The Eastern Orthodox Church* broke from Rome officially in the eleventh century. The Eastern Church represented Christendom, as established by the Emperor Constantine in the Eastern Roman Empire. It comprises the Greek, Russian, Armenian and Coptic Churches, independent Churches which are, however, linked together. While the Roman Church always claimed to be above temporal power, the Eastern Church has been closely linked to the state, often subservient to it. There are now between 100–150 million adherents.

(c) *The Protestant Churches* emerged in the sixteenth century consequent upon the Reformation; they developed from the reform movements which challenged the corruption and laxity of the Renaissance Papacy. Protestantism appealed to national feeling involving, as it did, a repudiation of foreign influences. Thus Protestantism adopted national forms and organisations, e.g. Lutheranism came to be a national Christianity for several of the German states and for the Scandinavian countries, as Anglicanism came to be for England and Calvinism for Switzerland, Holland and Scotland. Protestantism is divided into many denominations. The total number of adherents is over 250 million.

22. Judaism. The Jewish system of religious beliefs, practices and rites is known as Judaism. At an early date the Hebrews abandoned polytheism for monotheism. They linked morality with religion. In the first century A.D. the Jews were driven out of their homeland and scattered. The dispersion of the Jews, known as the diaspora, led to them seeking refuge in other lands and now they are to be found in almost every country in the world. They have been persecuted down the centuries, often with great cruelty, but their sense of community and unity has never been broken. Although numerically small—there are not more than about 15 millions of them—they have wielded an influence altogether out of proportion to their numbers. Zionism, the nineteenth- and twentieth-century movement to encourage the Jews to return to Palestine, ultimately resulted in the post-war creation of Israel and between 1948 and 1970 about 1,300,000 Jewish immigrants entered the country. Christianity is deeply rooted in Judaism.

23. Islam. The third great monotheistic faith is Islam, which sprang out of the teachings of Muhammad (A.D. 570–632), an Arab. During the two centuries after Muhammad's death Muhammadanism made great progress, sweeping across North Africa and Central Asia. Subsequently, it was carried down the east coast of Africa and across the Indian Ocean into Indonesia. There are two main sects, the Sunnis and the Shias, the former outnumbering the latter by almost ten to one. Altogether, there are about 600 million Muslims in the world. Islam has a traditional antagonism to Judaism.

24. Hinduism. Hinduism is confined almost entirely to the Indian sub-continent but its adherents number over 400 millions. It is a development of Brahmanism, the belief in one divine spirit. Early Brahmanism was influenced by Buddhism. The Hindus worship many gods and goddesses, although these are all regarded as being manifestations of Brahma. The religion preaches yoga (physical and mental discipline, i.e. meditation and the mortification of the body) through which spiritual peace and happiness can be attained. An outstanding feature of Hinduism was the caste system which formerly prevailed: this rigidly divided people into groups. India's social and economic backwardness can be at least partly attributed to this caste system. Many reformers have attacked this system and caste is gradually being undermined: indeed, "untouchability" has already been officially abolished and its practice in any form is now punishable. The caste system is so ingrained in India, however, that it will be a long time before its influence is totally eradicated.

25. Buddhism. Founded by Gautama, the Buddha, in the fifth century B.C., Buddhism is concerned mainly with salvation; it eschews all speculation about God and the universe. It was a great proselytising religion overrunning the Indian sub-continent and then taking hold of China and south-east Asia. After a millennium it declined in India which largely reverted to Brahmanism. There are two branches of Buddhism: Mahayana Buddhism found in Tibet, Mongolia, China, Korea and Japan, and Hinayana Buddhism found in Sri Lanka, Burma, Thailand, Cambodia and Vietnam. It is believed that there are about 200 million Buddhists in the world.

26. Confucianism and Taoism. These two religions, found in China, sprang from the teaching of Confucius and Lao-tzu who lived in the fifth and sixth centuries B.C. respectively. Confucianism was founded upon the practice of morality as exemplified in the lives and teachings of the early sages. Taoism taught that man must try to live in harmony with nature. When Buddhism came to China it became mingled with these two ancient religions. Thus in China there was a kind of religious ideology which blended all three faiths. The coming of Communism to China has greatly undermined the traditional religious beliefs.

27. The geopraphical distribution of religions. Religion, as has been said, knows no frontiers, but some religions have fairly distinctive areas of occurrence (*see* Fig. 9). The spread and the adoption of any religion has largely depended upon three factors:

(*a*) The attractiveness of the creed or system of ideas.

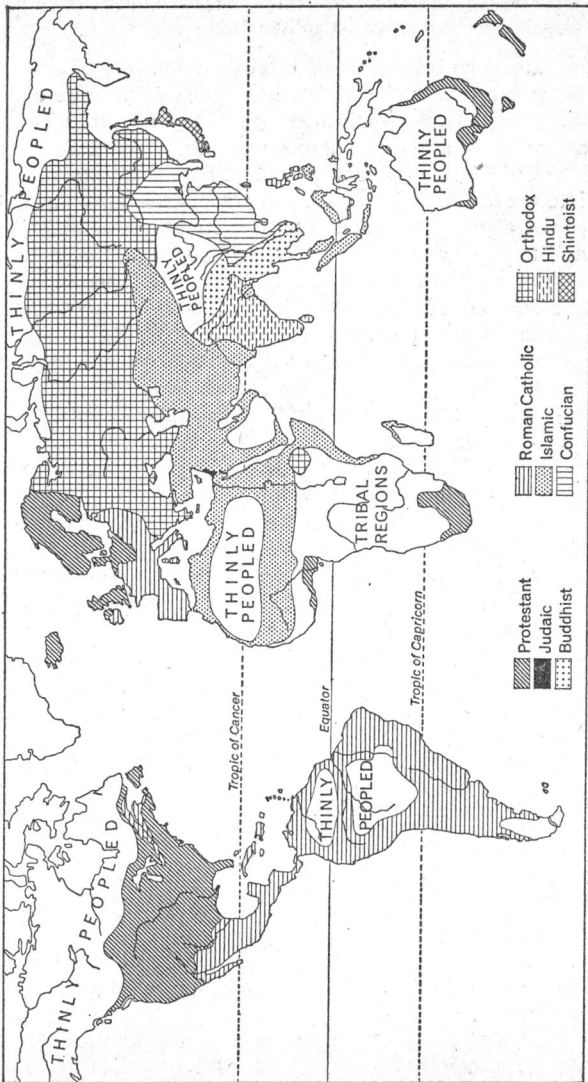

Legend:

- Protestant
- Judaic
- Buddhist
- Roman Catholic
- Islamic
- Confucian
- Orthodox
- Hindu
- Shintoist

FIG. 9 *The religions of mankind.*

The Old World is the home of the great religions of mankind and the Near East was the home of the three great monotheistic religions. Native peoples who are uncivilised are usually animists: that is, they believe that nature is peopled with spirits. The Christian religion with its various denominations has spread widely throughout the world. This map is, of course, greatly simplified.

(b) The vigour of the missionary zeal propagating the faith.

(c) The particular circumstances of the time.

However, in some cases one can discern the influence of geographical factors. The best illustration is probably the case of Islam. Attention has frequently been drawn to the correlation between its main area of distribution and the arid and semi-arid area of the Old World (although we should note that Islam has spread into wet equatorial regions). Islam was spread with unique rapidity by Arab warriors but came to a sudden halt when different forms of society and different ways of life were met with in western Europe, eastern Europe, China Proper and in the India of heavy monsoon rains. An interesting and outstanding fact is the persistence of the Christian faith in Ethiopia amidst the sea of Islam: one can point here to the very different geographical character of Ethiopia to that of most Islamic lands. (See H. J. Fleure, "The Geographical Distribution of the Major Religions," *Société Royale de Geographie d'Egypte*.)

Although the Christian religion made vigorous attempts to convert the East, the efforts met with ultimate failure. The expansion of Christianity beyond Europe was coincident with, and largely dependent upon, the outflow of Europeans across the oceans. As H. J. Randall says, "The reasons for this peculiarity of geographical limitation have never been properly examined. All that can be hinted here is that its absorption into the Mediterranean culture rendered it unfitted for export into lands where that culture was either exotic or absent" (*The Creative Centuries*, Longman).

28. The influence of geography upon religion. Every religion is modified or influenced to some degree by its surroundings.

(a) The objects of worship may bear some relationship to the environment:

(i) In countries where there is seasonal and uncertain rainfall the rain god is a common deity, particularly in India and parts of Africa.

(ii) In the high intermontane plateaux of the Andes, where it is cold except in the sunshine, sun worship prevailed amongst the Indians.

(iii) Amongst many coastal communities, who depended upon fishing for their livelihood, the god of the sea was of special significance.

(*iv*) The River Nile, "giver of life", was worshipped by the Ancient Egyptians, for in this arid desert land, the annual flooding of the river, which provided water for irrigation, meant the difference between life and death.

(*b*) The idea of heaven is largely conditioned by the environment:

(*i*) In hot, dry, parched lands heaven is a land of plentiful water, cool fountains and running water.

(*ii*) The Eskimo, denizen of the arctic wastelands, conjures up a heaven of warmth and plentiful food.

(*c*) Temples and representations of the gods may be influenced by the surroundings:

(*i*) The Christian cathedrals of the Middle Ages were built with soaring towers and spires. Also they were built with much window space to let in as much light as possible.

(*ii*) Hindu temples and images of the deities are elaborate and exotic and seem to reflect the exuberance of the natural vegetation, while the elephant is a common motif.

(*iii*) Islamic mosques have few wall apertures in order to exclude heat; but they have much mosaic work to catch and reflect light and to give coolness.

(*d*) Certain plants may become cult symbols in particular religions:

(*i*) The bo tree and the lotus plant in Buddhism.

(*ii*) The oak and spruce amongst the early Germanic tribes.

(*iii*) Coniferous trees in Shinto.

(*iv*) The mistletoe amongst the ancient British Druids.

(*e*) The imagery used in religious writings may reflect the environment: this is particularly true in the case of the Bible where the references to sheep and shepherds reflect what was a leading occupation in Biblical times.

29. The influence of religion. Religion exercises a much more profound influence upon social and economic life than one might imagine. Among societies where religion has a strong hold, its effects can be very powerful. Here are some examples of how religion may influence man's activities and outlook.

(*a*) The Roman Catholic Church forbids the use of contraceptives and this clearly is likely to affect the population increase in Catholic countries; e.g. it has been largely responsible for the demographic problems of Italy.

(*b*) Ancestor worship in China, polygamy among Muslim

peoples and early marriage among Hindus have all greatly influenced the distribution and density of population in particular
areas.

(c) Because the Jews, dwelling in foreign lands, were forbidden
to hold land or to engage in manufacture they turned to commerce and finance and so often became wealthy and influential.

(d) Islam has acted as a restrictive force from the economic
point of view, for such activities as mining, trading and money-
lending are forbidden to members of the faith.

(e) The Hindu caste system in many ways militated against
efficient economic development, e.g. by restricting particular
activities to particular castes.

(f) The Hindu reverence for the cow has burdened India with
a vast and virtually unproductive cattle population which also
forms a great drain on the food supply.

(g) Buddhism forbids the killing of animals and so pre-
supposes a vegetarian diet and a poorly developed animal
husbandry.

(h) The pig is thought to be an unclean animal amongst
Muslims and Jews; hence pig-rearing is absent as a farming
activity in Islamic countries and in Israel.

(i) Difference of religion may give rise to strong antagonisms
between peoples and create difficult political problems, e.g. between Muslims and Hindus, Jews and Arabs, the Protestant and
Catholic Irish.

GEOGRAPHY AND ARTISTIC EXPRESSION

30. The geographical factor. Geographical conditions have had a
direct effect upon man's cultural development, and art, music
and literature often reflect, though not always very obviously,
the influence of the environment.

As in the case of religion, artistic expression has geographical
relationships.

(a) The spread of artistic ideas is influenced by geographical
position and other natural conditions.

(b) The ideas, themes and content, whether in art, music or
literature, may be influenced by the environment.

31. Art. The impressions which man derives from his environment form one of the mainsprings of his art and it is the *genius
loci* which gives rise to artistic styles and ideals. Moreover,
especially in architecture, the raw materials at man's disposal

often indicate the form and character of his building. Early art tends to show the impress of local environment or the regional environment much more forcibly than contemporary art, since the widening of intercourse brought enlarged resources and increased the movement of ideas. The significance of the geographical factor is often best illustrated by the fact that imported artistic styles frequently become adapted, modified and transmuted into something new in a different physical and social environment.

This is not the place to discuss the influence of geographical factors upon the fine arts but a few examples will illustrate the point.

(*a*) The nature of the available local building material has formed the basis of regional architecture since the form, size, resistance and colour of that material invests buildings with a particular character; e.g. in alluvial plains where stone is absent brick is often used.

(*b*) Climate appears to have had some influence upon decoration in artistic design. In regions of strong sunlight where light and shade are accentuated, and sun and shadow produce sharp lines and shapes, the main tendency is towards geometrical design, e.g. geometric patterns in Islamic carving, Aztec mosaics and Inca textiles.

(*c*) The presence or absence of stone influences the degree to which sculpture becomes a developed art. The perfection of Greek sculpture is largely due to the use of marble, available from the Hymettic and Pentelic quarries near Athens, which is particularly suited to carving. In contrast, the Dutch, who have excelled in pictorial art, have never produced a sculptor of the first rank, the reason possibly being the dearth of stone in the Netherlands.

(*d*) The influence of animal and vegetable life is primarily upon decoration in art. Vegetation—trees, leaves, flowers—has formed the basis of much design. In India the architectural column displays a remarkable variety of form and ornament, reflecting the variety and exuberance of tropical vegetation. A notable feature of Hindu art is its close connection with animal life; the bull in early times and the elephant later play a very important role.

32. Literature. The geographical influence in literature shows itself chiefly in the settings or backgrounds of novels. English literature is particularly rich in the so-called topographical novel:

Trollope, the Brontës, Thomas Hardy and Arnold Bennett all used the local environment in which they lived as the backgrounds for their works, e.g. Hardy's Wessex, the Brontës' Yorkshire moors, Bennett's Five Towns. The environment has also had its impact upon the poets; Matthew Arnold wrote of Wordsworth: "It might seem that nature not only gave him the matter for his poem, but wrote the poem for him."

33. Music. The environmental influence in music is much more subtle and less obviously apparent but it is there nevertheless. It is probably most obvious in the compositions of people like Sibelius; he himself said that nature, was his inspiration. "Certainly the brooding vigour of his music evidences exceptional sensitivity to the natural harmonies and rhythms of nature. The stark pines, the discordant winds, the mists rising over the dark lakes, the sound of rain falling among the birch trees—these, and the legends in which Finland is so rich, are the stuff of his musical mind." (*Everyman's Encyclopaedia*, Vol. II, 1958, pp. 320–321.)

Some composers achieve a nationalist feeling in their music which is often obtained by deliberately borrowing or imitating the folk-music as, for example, Borodin in Russia, Smetana in Bohemia and Grieg in Norway.

CULTURE REALMS

34. The cultural region. Geographers recognise various types of region, for example, structural, climatic and natural. During more recent years, greater attention has been paid to the concept of the cultural region. This kind of region is one that is distinguished not by the physical conditions of the environment, though these may have some bearing upon its characterising features, but rather by a distinguishing set of cultural traits, i.e. distinctive languages, beliefs, customs, modes of behaviour, social institutions, ways of life, artifacts, etc. The sum total of all these things which, it must be remembered, are closely connected with man as an individual and men as a group, rather than with the habitat in which either the individual or the social group lives, produces a distinctive cultural entity.

A group of people speaking the same, or kindred, languages, practising the same religion, observing similar manners and customs, engaged in similar types of work and living in similar types of community groups—in other words unified by the same or very similar cultural traits—may be said to form a culture group. The geographical area occupied by such a culture group constitutes a cultural region or cultural realm.

35. The recognition of culture realms. How can one distinguish cultural realms in the world? By which criteria can we distinguish them? To turn to the first question, a major cultural region or realm, to be acceptable, must satisfy two conditions:

(*a*) There must be some aspect of its culture which pervades the area and such a combination of cultural features common to the area that it may be recognised as an entity.

(*b*) There must be fairly strongly differentiated cultural features between it and neighbouring areas to justify its recognition and to allow its boundaries to be demarcated.

The term culture region implies an area of relative cultural uniformity rather than one of absolute uniformity.

With respect to the second question regarding criteria, it has already been indicated in **34** that a very wide range of cultural traits exists and may be used to differentiate groups, but perhaps these can be reduced to certain criteria, notably language and literacy, religion and beliefs, social organisation, social outlook settlements and architecture and economic activities and organisation.

36. The major culture realms. Although geographers differ slightly with regard to the number of separate realms they distinguish, there is a fair measure of agreement over this and also over the boundaries of the culture realms. A simple grouping could be as follows:

(*a*) The Polar Realm.

(*b*) The European Realm.

(*c*) The Anglo–American Realm.

(*d*) The Latin American Realm.

(*e*) The "Dry" Realm.

(*f*) The African Realm.

(*g*) The Oriental Realm.

(*h*) The Australian–New Zealand Realm.

(*i*) The Pacific Realm.

(*j*) The Communist Realm.

Some geographers link (*b*), (*c*), (*d*) and (*h*) together as a single Occidental Realm and think of these different areas as subregions. The Oriental Realm may be divided into three sub-

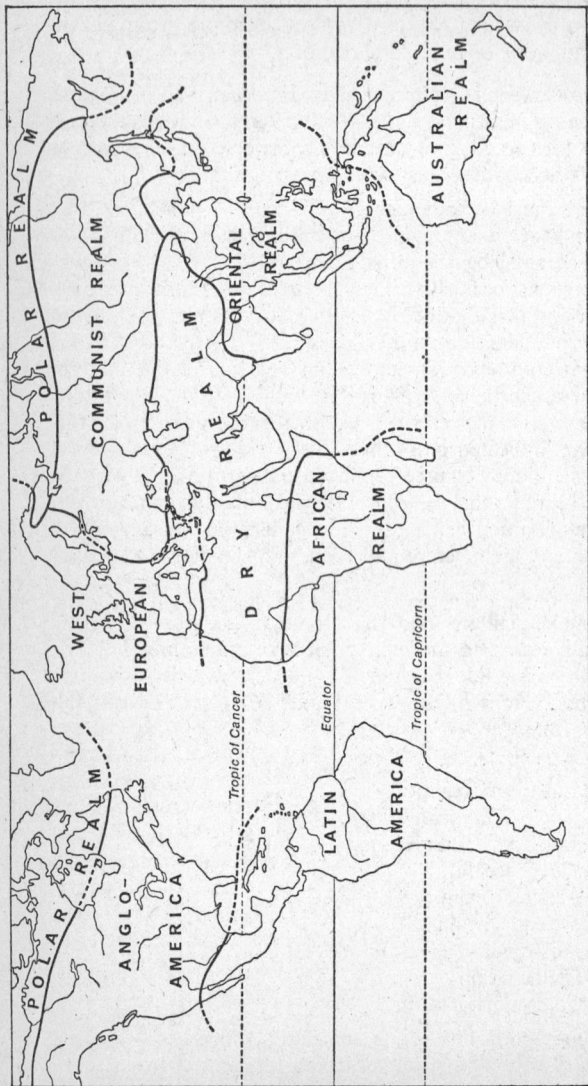

FIG. 10 *The culture realms of the world.*

This map shows all the main realms; each realm could be sub-divided into regions. The U.S.S.R. and the communist countries of eastern Europe might be called a separate culture realm because of their distinctive economic and political organisation.

regions: the Indic, Indo-Chinese and East Asian; although some authorities recognise these as major regions. Figure 10 shows the principal culture realms.

37. The Polar Realm. This is the high latitude world of perpetual snow and ice, tundra and the northern fringes of the taiga. Tribes of Mongoloid origins made their way into these polar lands at an early date. They developed a simple economy based very largely upon the exploitation of animal life: they became either hunters and fishers or herders (of reindeer). Agriculture was precluded because of the severity of the climatic conditions. Natural resources were extremely scarce and the best possible use had to be made of animal products—fur, hide, bone, horn, sinews, etc.

The inhabitants are partially nomadic. Their dwellings are simple and crude but often remarkably effective as shelters. Social distinctions based on wealth hardly exist and "economic relations verge on a primitive communism" (R. J. Russell and F. B. Kniffen, *Culture Worlds*, Macmillan). Political organisation has never developed, largely because the area is too sparsely populated. Architecture is absent and there is no art save for decoration and some carving of horn. Isolation and the unique culture traits of the polar territories set the area apart as a culture realm quite different from all the others, although gradually its identity is being lost as a result of contacts with, and influences from, other cultures.

38. The European Realm. European civilisation or Western culture springs from a Graeco–Roman Christian base and Europeans, generally, have been aggressive, acquisitive and progressive. Although there are ethnic differences, distinguishable as Alpine, Nordic and Mediterranean types, and there is great linguistic diversity, most Europeans are nominally Christians. Highly developed field agriculture of the mixed farming type, industrialisation and technological development, urbanisation and social mobility, a high degree of occupational specialisation and a rare artistic, musical and literary creativity are the outstanding characteristics of European culture.

Europe has been the birthplace of many important and highly influential political ideas such as democracy, nationalism and communism and the European Realm is distinguished by its acute political fragmentation. This fragmentation has, in the past, led to rivalries which often erupted into nationalistic wars.

The second World War produced irreparable damage to Europe and at the end of it the protagonists were left exhausted, shattered and economically bankrupt. From this disruption, however, co-operation both on the economic and political planes began to emerge which eventually led to the creation of the European Economic Community which has done much to rebuild western Europe.

If less influential politically than formerly, western Europe has regained its economic importance. For the most part, the peoples are literate and well-educated, healthy and energetic, and enjoy high standards of living. Notwithstanding the current economic recession, inflation and a serious measure of unemployment, largely contingent upon the energy crisis of 1973, Europe's influence is still to be reckoned with and she forms a third force in the world after the United States and the Soviet Union.

39. The Anglo-American Realm. The cultural characteristics of North America north of the Rio Grande are fundamentally derived from Europe but mostly they seem to have become over-emphasised. Most things are brash and grandiose, and the muted tones are few. The prolific richness of resources and the newness, rawness and spaciousness of the environment have had a great impact upon the people.

Canada and the United States were originally colonised by the British and the French but during the nineteenth and early twentieth centuries large numbers of Irish, Italian and Central Europeans migrated to North America. Although multinational in origin, the various groups in the United States soon became assimilated and their allegiance to the land that had welcomed them soon became firmly established. Assimilation in Canada was less effective and we are all well aware of the separatist movement of the French Canadians who have preserved their identity. However, the British influence is paramount as the very term Anglo-America suggests.

Capitalism, industrialism and urbanism here reach their peak. The economic exploitation of resources has been ruthless, farming and manufacture are highly mechanised and often automated, output from soil and mine and factory is prodigious and the result is the highest standard of living in the world (though locally there are exceptions). The people are literate and well educated, vocal and aggressive. Some American culture features, e.g. the modern-type city with its glass and concrete skyscraper features, have spread widely throughout the world. Indeed, occidental culture in its Americanised form is insidiously spreading into almost every corner of the world.

40. The Latin American Realm. Whereas the English language and institutions have moulded the predominant culture pattern in Anglo-America, Iberian tongues and institutions have been dominant in Latin America. The Roman Catholic Church is still strongly entrenched, and many cultural features associated with land-holding and law have been imported from Mediterranean Europe and still prevail. Mediterranean architectural forms are common in many villages and small towns but in the larger cities American influence is strong. Native Indian cultural influences, e.g. in artistic design, are apparent in many areas in the region. Ethnic diversity but with a marked degree of ethnic fusion also is characteristic.

Socially the region is distinguished by the wide gap between the few very rich and the many very poor, by the generally low standards of living, and by the high degree of illiteracy. Political instability has been a marked feature of the region with a tendency towards dictatorial government. but now rather more stable political conditions seem to be prevailing.

41. The "Dry" Realm. Running obliquely across the heart of the Old World land mass is an extensive belt of arid and semi-arid climate which has given rise to desert and semi-arid grassland. This "dry world" has formed a significant barrier and divides the peoples of the culture realms on either side of it. Moreover, it has developed a cultural distinctiveness and unity of its own. Formerly, most of the peoples were typically nomadic and largely dependent upon herding, although where plentiful water supplies occurred, supplied by rivers and wells, sedentary agriculture was practised. Cultivation was more or less restricted to oasis-type settlements. Historically, the carrying on of trade was closely associated with herding, and caravan routes linked the oasis villages and towns. Society, largely tribally organised and nomadic, and mainly Muslim, made the dry world a relatively homogeneous culture region, although two wings, the Arab–Berber wing in the west and the Turko–Mongolian wing in the east, were traditionally recognised. Ethnically and linguistically there is no great homogeneity, rather a great diversity, but the Islamic faith, if not common to the whole realm, is widely spread and is a strong cementing factor.

The discovery and exploitation of rich oil deposits, particularly in the Middle East but also in North Africa, has radically altered the economies of many of the hitherto poverty-stricken countries. In many of these oil-rich countries, e.g. Saudi Arabia, the Persian Gulf sheikdoms, Iraq and Iran, rapid transformations are taking place, although it is still true to say that the wealth and influence lies in the hands of the few, while the many remain relatively underprivileged. The vast wealth accruing from oil revenues is making it possible for some countries to modernise and to develop their economies. Iran has made most progress so far.

Within this large culture realm lies the small state of Israel, the home of the Jews. In almost every way they differ from their Arab neighbours and Israel has become the focus of Arab hatred and a political danger spot in the Middle East.

42. The African Realm. Major ethnic, linguistic and religious boundaries, which also roughly coincide with the boundary between the desert and the savanna, separate the "Dry" Realm from the African Realm. The latter is the land of the Negroes although in the southern parts many white men have settled and made it their homeland, but always the whites form minority groups in the population. Measured by European standards, most African culture is relatively primitive. Although Bantu is fairly widely spread, there is sufficient linguistic diversity within the region to make inter-tribal communication difficult. Furthermore, the Negroes failed to develop written languages. Their building has always been simple and advanced architectural forms rarely materialised; their technological development was limited.

Negro economic development was, and for that matter still remains, simple: the people generally, except perhaps where they have come under the white man's influence, are either simple food gatherers and hunters or pastoralists or hoe cultivators. Society was essentially tribal, and political organisation of an advanced kind never emerged. Religion is often animistic and superstition and witchcraft are still fairly widespread among the more backward groups.

The coming of white men to Africa has had a big impact upon Negro society and yet, notwithstanding this, the majority of Africans are still basically poor, backward, illiterate, and often plagued with ailments and diseases. European control was not always beneficial, for slavery, forced labour and detribalisation took place and economic development was essentially exploitive.

Independence from European control after the Second World War provided many of the better endowed countries with opportunities for development and where they came under capable leadership they are making great strides; this is especially true of Nigeria and to a lesser extent of Kenya, Zambia, Zaïre and Ghana. Unfortunately the early democratic forms of government which were often tried have fallen prey to dictatorial military rule, e.g. Ethiopia, Chad, Uganda and the Central African Empire.

43. The Oriental Realm. This realm, which is broadly coterminous with Monsoon Asia, is a region of ethnic, linguistic and religious diversity and complexity and one is, therefore, tempted to ask: wherein lies the common cultural denominator in all this rich diversity? Briefly, it is a socio-economic denominator for with a few exceptions (notably Japan, Hong Kong, Singapore) material poverty, economic inertia, natural fecundity, illiteracy and the rural life are characteristic of much of the area. Professor Dobby, writing about the realm a quarter of a century ago, said: "Monsoon Asia is the home of a 'vegetable civilisation', where wood and plant materials have been the basis of everyday things from houses to domestic utensils, farm implements and clothing" (*Monsoon Asia*). Although there have been changes, Dobby's statement continues to have a strong element of relevance to the realm as a whole. In many parts of the Oriental Realm, more especially in the Indian sub-continent and in much of south-east Asia and Indonesia, existence is at a miserably low level and poverty is the norm. In the past, high fertility rates, high mortality rates, and a low expectation of life at birth were characteristic, but these demographic features are changing for the better. The preponderantly rural sector of the population traditionally has been static, hidebound and conservative with little will and very few opportunities to effect the changes necessary in the economy to improve the lot of the population. But, in some areas, notably in eastern Asia, there have been in recent years quite revolutionary changes.

The spread of Communism to almost all the mainland countries of eastern and south-eastern Asia is likely to cause radical changes as has happened in China where life and activity in all its forms have been revolutionised. During the past twenty-five years the traditional peasant agriculture of China has been changed out of all recognition, as has social life, while the manufacturing industry has been greatly developed and widely spread. So rapid and so fundamental have been the changes that China can no longer be counted among the developing countries. Indeed, she is rapidly emerging as a world power.

Japan, highly industrialised and urbanised, enjoys a lucrative world-wide trade which has made her very prosperous. And, unlike most of the other countries, her former rapidly growing population has become stabilised. It is fair to say that modern Japan has more in common with the developed countries of the West than with her Oriental neighbours.

Within the major cultural region there are appreciable differences and diversities and these permit a three- or four-fold subdivision: the Indian region, the Far Eastern region, the Indo-Chinese region and the East Indian region (the last two are sometimes combined).

44. The Australian–New Zealand Realm. Were it not for its physical apartness, this area would not be distinguished as a separate realm; it is really, like Anglo-America, an off-shoot from Europe and part of a great Occidental Realm. Sometimes it is included in the Pacific Realm, but this is not very satisfactory. What distinguishes the Australian–New Zealand Realm from its neighbours is its European, and more particularly its British derived, culture. The aboriginal peoples of both Australia and New Zealand were too few in number to have exerted more than the slightest influence upon the European incomers.

Typically the region is one of balanced agriculture and industry. Living standards are high, the peoples are well educated, energetic and progressive while there is a high degree of urbanisation. American cultural influences are growing.

45. The Pacific Realm. The island world of the south Pacific may be regarded as a separate culture realm, but three sub-regions are usually distinguished: Melanesia, Micronesia and Polynesia. The population of the realm is only just 5 million but the area has a cultural diversity as varied as any other in the world. Some of the most primitive peoples on the face of the earth inhabit these islands yet their "cultural development was unique in many ways. Their material wealth was small and they possessed few cultural heritages of Oriental World origin. They developed among themselves some of the most original social organisations known to the modern world" (R. J. Russell and F. B. Kniffen, *Culture World*).

In Oceania, as might be expected, the maritime environment has had a great influence: the sea has been a source of food, a great highway and the source of much native mythology. The economy is of the subsistence type involving collecting, rudimentary tillage and fishing, except where in some instances, European colonisation and control has led to some mining and some plantation agriculture.

But there is a ferment of change in Oceania arising partly from the effects of *papalagi* (i.e. foreign) culture and partly from a native cultural awakening. As Albert Wendt has written: "There is a multiplicity of social, economic, and political systems all undergoing different stages of decolonisation, ranging from politically independent nations (Western Samoa, Fiji, Papua New Guinea, Tonga, Nauru) through self-governing ones (the Solomons, the Gilberts, Tuvalu) and colonies (mainly French and American)." (*Courier*, February 1976, p. 10.)

46. The Communist Realm. Eastern Europe and the greater part of the Soviet Union are culturally closely akin in many ways to western Europe. On the other hand, the Soviet Union and the Communist countries of eastern Europe may lay some claim to be considered as a separate cultural region because of their distinctive economic and political organisation. Communism, based on the teachings of Karl Marx, supplies the basis for social and economic organisation. Capitalism and democracy of the liberal kind developed in western Europe is anathema to the Communists although the late 1970s has shown some weakening of the unified front adopted by the U.S.S.R. and its satellite states in the organisation of internal affairs along rigid anti-capitalist lines. The Communist Party is the driving force behind the political machine. The Soviet Union is of particular interest, however, since it was the first country to adopt planning of the national economy, planning which, it must be allowed, has been remarkably successful. Indeed, most countries these days have followed suit and pursue planning policies to a greater or lesser degree.

PROGRESS TEST 4

1. Attempt to define the term "culture". (1)
2. Why is culture said to be dynamic? (2)

3. Quote four examples to illustrate how cultural features have spread from one part of the earth to another.   (7)

4. Explain the meaning of the terms: determinism, possibilism, probabilism.   (9)

5. "Man has gained an increasing ascendancy over nature." Do you agree?   (10)

6. Indicate some ways in which man has changed the natural landscape.   (11, 12)

7. Why is language important to man?   (13, 16)

8. Name the most important languages of commerce. What is a lingua franca?   (17, 18)

9. Describe the geographical distribution of the chief religions in the world.   (20–27)

10. Quote examples to show how geographical conditions have influenced religion.   (28)

11. In what ways may religion influence human activities?   (29)

12. Discuss the relationship between art and environment.   (30, 31)

13. Explain what is meant by the term "cultural region".   (34)

14. Describe the outstanding cultural features of *either* the African culture realm *or* the Latin American culture realm.   (40, 42)

CHAPTER V

Food and Health

HUMAN HEALTH

1. Physical health. Some peoples are more robust and energetic than others, but these features are due rather to general health, food and climate than to any innate racial differences. Many Orientals, especially the peasant rice cultivators, such as the Chinese, Filipinos and Indonesians toil laboriously for long hours ploughing the muddy rice fields and transplanting rice shoots, even though they are but slightly built and inadequately fed. The Malays and the South American Indians are sometimes said to be lazy but this is probably due to the climatic and other conditions under which they live rather than to any inferiority in their physical make-up. Europeans, the peoples of Canada and the United States, Australians and New Zealanders, in general, enjoy better health than most peoples for these reasons:

(*a*) The climatic conditions under which they live are more conducive to healthy living.

(*b*) Their diet is better and more balanced and this helps to keep them fit and to ward off certain ailments.

(*c*) Conditions of hygiene are superior and medical services (doctors, clinics, hospitals, etc.) are available.

2. Race and health. The charge made in earlier times that some races are naturally physically and mentally inferior is invalidated by impartial investigation: if the physical health of particular groups of people is generally poor then the condition can be blamed upon such factors as food supply and living conditions and not upon any constitutional inferiority.

On the other hand, it can scarcely be doubted that some peoples are better adapted to specific environmental conditions than others; the Negro, for example, is better adjusted to the heat and humidity of tropical regions than the white man, and accordingly he lives more comfortably and is less affected by the climatic conditions.

3. Optimum climate for man. Many years ago the American geographer Ellsworth Huntingdon put forward the suggestion that cool, temperate climates with their distinct seasonal, even daily, changes in the weather conditions and lack of extremes of temperature offered the best conditions for a maximum output of energy. He also believed that such climatic change acted as a mental stimulus. Certainly the most notable advance in modern science, technology and the arts have come from peoples living in such environments. Huntingdon also pointed out that, historically, the various civilisations of Egypt, Greece, Rome, France, and the modern industrial civilisation which began in England have moved progressively poleward, i.e. from subtropical into more temperate latitudes. In contrast, tropical environments tend to be enervating and are not conducive to strenuous physical activity. On the other hand, the extreme cold in winter of many high latitude lands and areas in continental interiors may make outside work well-nigh impossible.

4. Disease. Man is subject to many ailments and diseases: some of these, such as the common cold, are merely a temporary inconvenience but others, such as cancer, may be fatal. From time to time in the past, epidemics and plagues such as the Black Death in the Middle Ages and the great influenza epidemic of 1918, caused large-scale loss of life. The chances nowadays of such occurrences have been substantially lessened by isolation, hospitalisation, inoculation, etc., but the world is still troubled by many medical problems.

One of the interesting facts that emerges from the study of diseases is that the prevalent and serious diseases of one age are replaced by newer diseases of another age. Smallpox, which was once a scourge in England, now seldom occurs; again tuberculosis, formerly widespread, has been more or less eliminated. By systematically attacking the breeding grounds of the anopheles mosquito, the carrier of malaria, this disease has been virtually eliminated in Sri Lanka. Although man can claim many successes in his fight against disease, there are still many serious medical problems waiting to be solved such as cancer and the new nervous disorders which seem to be more especially the outcome of modern ways of life.

5. Some tropical diseases. Although some diseases, such as yellow fever, may be serious and fatal, many of the most troublesome ailments are those which are not immediate killers but which are insidious, maiming and debilitating. Diseases are more rife in tropical environments than elsewhere. Some of the more common diseases are the following:

(a) *Malaria.* Seldom a killing disease amongst native peoples, malaria is, however, a great weakening disease which ultimately may lead to death. Although still rife in tropical regions generally, great inroads have been made against it and many areas practically cleared of it.

(b) *Yellow fever.* This is endemic throughout most of tropical South America and Africa. It can now be controlled by vaccination and is much less of a problem than it used to be.

(c) *Cholera.* An infectious bacterial disease which formerly had an almost world-wide occurrence and accounted for many deaths. It is now largely confined to Monsoon Asia where it is endemic, but occasionally outbreaks occur elsewhere, as recently happened in Portugal, Spain and North Africa.

(d) *Dysentery.* This is mostly a disease of tropical regions. The fatal effects of this, and of many other bacterial diseases, have been substantially diminished by the use of modern drugs and antibiotics.

(e) *Hookworm.* This and other worm infestations are more particularly associated with hot, wet tropical climates. The practice of geophagy (earth eating) combined with the fact that the natives often do not use footwear causes infection.

(f) *Influenza.* This is a virus disease. Europeans have developed a high degree of immunity to influenza unless it takes a virulent form but in many parts of the world in the past it has been a fatal disease.

(g) *Kwashiorkor.* This is a widespread malnutritional disease, particularly prevalent in tropical lands where the diet is largely one of carbohydrates and protein foods are in short supply. It is very common amongst young children and leads to body swelling and physical malformation.

(h) *Leprosy.* A chronic transmissible disease which was a terrible scourge in the past. Although it gradually disappeared from Europe after the fifteenth century, it is still found in Africa, south-east Asia and the Pacific Islands. It is estimated that the number of infected individuals may be as high as 10 million.

(i) *Smallpox.* This is a disease now known to be caused by a virus but there are several varieties of this highly contagious disease. It is characterised by fever and spots which leave behind disfiguring pock marks. Since the introduction of vaccines, smallpox has been more or less eliminated.

(*j*) *Trachoma*. This is a form of blindness, mainly induced by lack of cleanliness; it can now be cured by sulfonic drugs as well as some of the antibiotics. Even so it is very common in the Middle East and parts of Africa and the number afflicted runs into millions.

(*k*) *Bilharzia*. This is a disease caused by infection from a fluke parasite, the schisosome; it leads to the discharge of blood and is very weakening. It is prevalent in Africa, the West Indies, tropical South America and in parts of south-east Asia.

6. Causes of disease. Let us now try to summarise the causes of disease. There are many, of course, but broadly the chief causes are listed below:

(*a*) Inadequate food supplies causing malnutrition.

(*b*) Contaminated water and the paucity of treated water supplies.

(*c*) Over-crowding and unhygienic living conditions.

(*d*) The lack of sanitation and sewage disposal systems.

(*e*) The prevalence of insect pests and vermin which are carriers of diseases.

Many of the people in tropical regions are poor, illiterate and have an unbalanced diet. They may sometimes pay scant attention to elementary hygienic precautions and are thus prone to contagious diseases which can spread with astonishing rapidity.

7. Principal problems. There are many problems to be tackled and solved in respect to the medical situation but perhaps the three principal problems facing authorities are as follows:

(*a*) The general lack of trained doctors, nurses and other personnel concerned with health.

(*b*) The often acute shortage of hospitals, clinics and medical equipment.

(*c*) The difficulty of overcoming tradition, superstition and ignorance connected with medicine and health which prevail amongst the great mass of the population.

It will be appreciated that these three problems are inescapably linked with two other problems: namely, the high degree of illiteracy and the inadequacy of education and training facilities, and the national poverty and the lack of funds for establishing hospitals and training medical personnel. Hence, there is a vicious circle.

8. Solutions. The problems of medical geography clearly require money and trained specialists but, above all, the problem needs to be tackled at its root. Education is the essential prerequisite for over-all improvement. It is instructive to compare maps showing the degree of literacy with other maps giving the incidence of certain diseases, infant mortality and near starvation: they show some remarkable similarities. Not all the problems and difficulties can be defined in terms of illiteracy and education, but it is certain that literacy would help solve them. Literacy provides the basis for material progress and all it implies—scientific and technical advances, improved hygiene and health, social betterment, etc. Moreover, education is the best antidote to the belief in social inequality, superstition and stifling tradition.

POPULATION AND FOOD SUPPLIES

9. Food and population. The study of population, human health and disease is inescapably linked with the complementary study of food production and resources. The world's population, as has already been indicated, is expanding at such a rapid and alarming rate as to warrant it being termed an "explosion". Many intelligent people believe that this very rapid growth is *the* most important and most pressing problem of our time and the one fraught with the most serious consequences.

The problem which this growth presents falls into two parts:

(*a*) Whether there will be enough living space for the population of the future.

(*b*) Whether the world will be able to feed adequately the additional population.

10. Living space. The problem of living space is the least troublesome of the two. From a purely quantitative point of view, it has been calculated that the world's present population—around 4,000 millions—could be found standing room on the Isle of Wight! Clearly, then, there is still plenty of room on this planet although, of course, mankind would not want to live like battery hens. Locally, where there is a shortage of living space, as in many of our big cities, man is now building vertically instead of horizontally.

Architectural engineers maintain that, with modern technology, it is structurally feasible to build skyscrapers five kilometres high capable of accommodating 30,000 people. If necessary, it is very likely that man would expand over the water, building vast rafts upon which he would live permanently in much the same way as the floating populations of Canton live at the present time, or would even resort to building undersea cities in gigantic undersea "bubbles". In case these ideas seem far-fetched and impractical a "sea city" has already been designed which is capable of housing 30,000 people.

11. Food supply. The problem of ensuring food supplies is much more serious. The threat of over-population and a shortage of food has been raised in the past, e.g. by Thomas Malthus at the beginning of the nineteenth century. But by the end of that century this spectre had been laid to rest as a result of the opening up of the "new lands" of the world which supplied grain and meat in large quantities.

In recent years the threat has begun to loom large once again and there can be no denying that the possibility of an over-all world shortage of food is very real indeed. The problem of finding sufficient food for the growing population of the future is aggravated, moreover, by the fact that already some two-thirds of the present world population are underfed, which means that they are suffering from either undernourishment or malnutrition.

The question thus arises: how can the world's food supplies be expanded to meet a growth in world population of over 100 millions a year?

12. Limitations of the land surface. In attempting to answer this question it will be useful first to review the world's productive and unproductive areas.

Only 29 per cent of the earth's surface is dry land. Furthermore, only a very small proportion of this is fertile and capable of being used agriculturally. Figure 11 shows the vast unproductive areas of the world, areas which are either of very restricted usefulness or completely useless.

The conditions which limit the use of the earth's land surface for cultivation may be summarised as follows:

(*a*) Approximately one-fifth of the total land area is *too cold*.

(*b*) Approximately one-fifth of the total land area is *too dry*.

(*c*) Approximately one-fifth of the total land area is *too mountainous*.

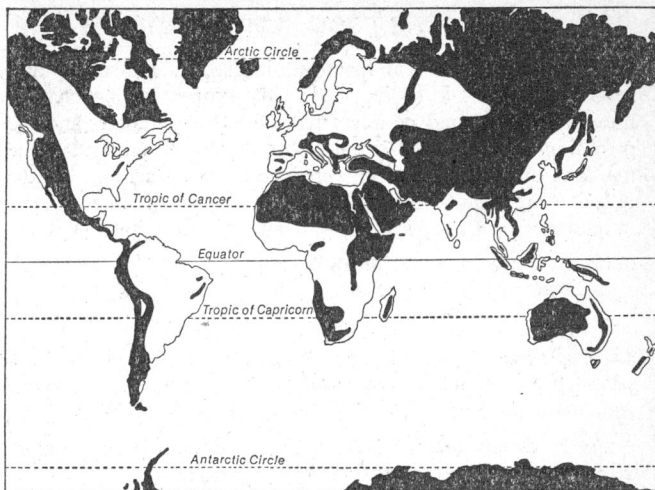

FIG. 11 *The world's productive and unproductive areas.*

Those parts of the earth's surface which through being too cold, arid or mountainous make close settlement and agricultural development well-nigh impossible are marked in black: these are the unproductive or "negative" areas. The blank areas, totalling only about one-third of the world's land surface, may be termed the productive or "positive" areas, i.e. those areas which are cultivated or are potentially cultivable.

(d) Approximately one-fifth of the total land area is *forested* or *marshy*.

Of the one-fifth (20 per cent) of the earth's surface that is available for agricultural use, only half of it (i.e. 10 per cent of the total land area) is actually producing food at present.

NOTE: In connection with this there are a number of points we might observe: (*i*) The best land is already being used. (*ii*) Much of the farming is below maximum efficiency. (*iii*) Some of the land which is not being used does offer possibilities for cultivation. A proportion, possibly as much as a quarter, of the 80 per cent of the earth's surface that is too cold, or arid, or high or forested could be made suitable for farming purposes, but only through a large capital investment.

MEETING THE FOOD SUPPLY PROBLEM

13. Increasing supplies of food. Additional, indeed substantial, supplies of food will be required in the future. The problem of present hunger could be alleviated, perhaps overcome, and the provision of extra food supplies for the future ensured if the methods of producing food were approached scientifically, the output shared equitably and the wastage of food eliminated.

Increased food supplies can be obtained in three main ways:

(a) By the extension of cultivable land.

(b) By increasing agricultural yields.

(c) By developing new sources of food.

14. The extension of food-producing acreages. The world's food-producing area could be increased quite substantially by bringing more land into cultivation and this could be done by these means:

(a) Terracing some of the steep slopes in mountainous regions which have suitable climates for crop growing.

(b) Extending and improving irrigation facilities in some of the semi-arid and arid areas of the world.

(c) Draining swamplands and improving flood-control measures in certain wet and low-lying areas.

(d) Clearing some forested areas, especially tropical forest areas.

15. Increasing agricultural yields. Although the additional acreages resulting from the above measures would be valuable, in total they would increase the cultivable area by only a relatively small amount. Moreover, the effort would be costly in time and money. Making "two blades of grass grow where one grew before" seems to offer a better prospect for solving the problem; in other words, increasing yields rather than increasing areas would appear to produce the more significant results.

Increased production along these lines could be achieved by the following methods:

(a) Increasing the use of fertilisers.

(b) Using better seeds and improving animal strains.

(c) Introducing and improving crop rotation.

(d) Extending the use of insecticides to control pests.

(e) Using more efficient farm tools and machinery.

(f) Controlling soil erosion and adopting soil conservation methods.

(g) Developing educational training for improved farming.

(h) Improving health and thereby labour efficiency.

All this is not meant to imply that nothing has been done along these lines: much, indeed, has already been achieved, although a great deal more remains.

16. New sources and methods of food production. If the rapidly growing population of the world is to be adequately fed there must be not only an increase in the total amount of food produced but also an increase in the quantity of protein food of which there is a gross deficiency. The main product of agriculture is carbohydrate; what is most urgently required is protein food, for the conventional bulk foods (cereals, potatoes, yams) are deficient in protein content. More protein foods such as meat, milk, cheese, eggs and fish are required to remedy undernourishment and malnutrition. These could be obtained by methods such as these:

(a) More intensive fishing.

(b) Cropping wild animals.

(c) Processing vegetable materials.

The ways and means by which man could increase the output of food and tap new sources of supply are discussed in **17-21** below.

17. Fisheries. There are many ocean areas rich in fish that are not adequately exploited, e.g. the South African and Australian areas. Even so, a word of caution is necessary: the fish of the sea are not unlimited in quantity and it is probable that the total catch could be increased only twofold, or at most threefold, without serious danger of causing extinction. In some areas, such as the North Sea, over-fishing has already occurred and the fishery is showing smaller catches already. Indeed, so serious has this become that a temporary ban upon herring fishing has been imposed in the North Sea. In the world at large, the depletion of whale stocks has led to restrictions in catches and a closed season, in an attempt to preserve the species.

"Fish farming", i.e. the breeding of fish in ponds, lakes and canals, has long been practised in the Far East; this practice could be greatly extended and, in fact, certain developments along these lines are now taking place in Africa where the great natural and artificial lakes (Victoria Nyanza, Lake Kariba) have been stocked with fish. The time is fast approaching when man must stop hunting fish and begin to farm fish.

18. Cropping wild animals. Another possible source of food is by cropping wild animals. The vast savanna lands of Africa are the haunt of large numbers of herbivorous creatures, e.g. the antelope, the eland, the gnu, the zebra and a scheme is being studied whereby the more thorough protection of these animals could be combined with their provision of additional food supplies. The eland, for example, is a docile animal which provides good meat and rich milk; it is more resistant to insect pests and disease than cattle; hence, it could be a most useful source of protein food. Conservation of other wild life could similarly be linked with food production. The capybara and the freshwater manatee might provide unorthodox sources of meat, and since they feed on water weeds and swamp plants, they would not compete for food with other land animals.

19. Processing of vegetable materials. Many plants are used as sources of vegetable oil: the seeds or fruits of the sunflower, the cotton plant, groundnut and soya plants all yield valuable oil. The vegetable residue left after the oil has been expressed is used either for making animal foodstuffs or organic fertiliser or is simply discarded as waste. This residue, however, is rich in protein and is potentially valuable as a human foodstuff, although at present it is frequently rendered unfit for human consumption through overheating during the oil expressing process or because it is contaminated. If it could be made edible, it is estimated that it could yield 20 million tonnes of protein (roughly double the world's estimated protein deficit).

20. Synthetic foodstuffs. Synthetic animal foods produced from natural gas, air and water have been developed by I.C.I. scientists at Billingham and it is estimated that the food is likely to be available in commercial quantities in the near future. The first trials with animals have proved successful. This is a major breakthrough with respect to the problem of animal feeding; moreover, it would cut Britain's bill for millions of tonnes of cattle protein rations. If man-made substitute foods can be successfully developed for animals, the next logical step would seem to be the production of similar food substitutes for man. The development of TVP (textured vegetable protein) from soya beans was the first development in this direction; since then a number of synthetic foodstuffs have been produced and some are now being produced in commercial quantities and marketed.

21. Hydroponics. This is a method of growing plants by water-culture or soilless methods. Broadly, the method consists of growing crops in tanks filled with a soil substitute and water which is charged with dissolved plant nutrients. The plants use the inert seedbed as an anchorage, growing upwards in the normal manner, and send their roots downwards to feed in the chemical solution. The success of hydroponics depends upon adequate sunshine, aeration of the roots and the appropriate solution for the needs of the crop. Capital costs are high and the process is, at present, too expensive for general use but it can be of use in exceptional circumstances as in certain arid, rainless areas; for example, the method has been used on a small scale in the Persian Gulf state of Kuwait, and it has also been tried in California. More general use of hydroponics as a method of cultivation has not yet developed; although it would appear to have some potential for expensive market garden crops, it seems unlikely that as a method it will seriously challenge normal methods of cultivation.

22. Improving the distribution of food. In spite of the overall world shortage of food, we frequently read of food being wasted or of surpluses being stored away. For example, at present large quantities of grain are in store in North America and there is a "butter mountain" in the E.E.C., while in the past there have been gluts of coffee in Brazil, of fish in Norway, and over-production of sugar in the West Indies. These products remain unsaleable because their producers are unable to dispose of them at a satisfactory price. Hence it is important that the purchasing power of the poorer peoples, most of whom live in Asia, Africa and Latin America, should be raised to enable them to buy food. They need also to be taught about the ways in which they can vary their diet so that they can live healthier lives.

One way of resolving the dilemma is for the richer and more advanced countries to provide financial and other aid to the underdeveloped countries so that they can buy more food and also improve their own agriculture and establish mining and manufacturing industries of their own. This is already being done, of course, to some extent, but it remains true that the rate of population growth in backward countries is only just being balanced by their capacity to raise their nutritional and financial standards, even with help from other countries.

23. Summary. To summarise, it may be said that the problem of increasing food supplies can be tackled in two main ways:

(*a*) By making more effective use of the land and greater use of the sea and inland waters.

(*b*) By using new and different kinds of food, perhaps even man-made synthetic foodstuffs.

The problem of feeding coming generations satisfactorily is not intractable. If all the world's farmers were able to raise their standards of production to the level attained by most Dutch farmers, the present world area which is now devoted to agriculture could support many times its present population. Progress is being made, but a quickening in its rate is urgently needed, for at present it would seem that population growth is outstripping food production. There is a clear threat of global hunger.

PROGRESS TEST 5

1. Which factors influence physical health and how does climate affect health? **(1, 3)**

2. What relationship, if any, exists between malnutrition, poverty and disease? **(4, 5, 7, 11)**

3. Which are the principal causes of disease? Name six tropical diseases. **(5, 6)**

4. Explain the meaning of the following terms: geophagy, kwashiorkor, hydroponics. **(5, 21)**

5. Discuss the chief problems in the fight against disease. **(6, 7)**

6. How could man overcome the problem of finding enough living space for the possible excess populations of the future? **(10)**

7. "Two-thirds of the present world population are underfed." Why is this? **(11, 12)**

8. In what ways can man increase (*a*) the area of productive land, and (*b*) the quantity of food supply? **(13, 14)**

9. The easiest way of increasing food supplies is to increase agricultural yields. How can this be done? **(15)**

10. In what ways could (*a*) fisheries and (*b*) wild animals help to increase food supplies? **(17, 18)**

CHAPTER VI

Resources and Economy

THE EARTH'S RESOURCES

1. The natural environment. Man lives in a bio-physical environment which comprises two components:

(*a*) The physical environment of rock structure, relief, water and climate.

(*b*) The biological environment of plant and animal life in all its forms.

In order to live (in the sense of to survive) in the natural environment man must derive from it certain basic human needs: air, light, water, food, warmth, clothing and shelter. From these he can sustain life. In the beginning primitive man sustained life at the lowest level; he just managed to endure life. Today, most people desire to do something more than merely endure life; they want to enjoy it. Hence, in addition to procuring the basic necessities for existence, man wants some luxuries, e.g. motor cars, gold watches, fine paintings, exotic holidays, etc.

All human requirements, no matter whether they be absolute necessities or frivolous luxuries, have to come from the earth's natural resources: air, sunlight, water, land, soil, vegetation, animal life and minerals. These he uses to satisfy his many needs. Basically, therefore, it is these natural resources which can be turned into useful products by the skill of man that constitute the real wealth of the world.

2. Kinds of natural resources. The natural resources which are available for man's use fall into two main categories:

(*a*) Organic or living resources: these include forests, natural pastures, wild life, fish and other marine life.

(*b*) Inorganic or non-living resources: these include air, water, mineral fuels, metals, non-metallic minerals and building stones.

NOTE: Soil, which is composed of both organic and inorganic matter, falls between the two groups, but it is a resource of fundamental importance.

3. Latent resources. Before we proceed to look more closely at these various resources one point needs emphasising. Although nature has provided a wide range of natural resources for human use, such resources do not become of significance or value to man until he has reached a particular stage of cultural development. The resources are present, but they remain hidden or unused until man requires them and can use them. Here are some examples:

(*a*) *Metals.* Primitive man used stones and flints for weapons. He could not begin to make his weapons and tools of metal until he had learned various skills and techniques:

(*i*) How to distinguish metallic ores from ordinary rock.

(*ii*) The techniques of smelting ores to extract the metals.

(*b*) *Rubber.* Man has known of the latex of the rubber tree for a long time, but rubber had little utility until there were certain developments in technology:

(*i*) The method of vulcanisation was discovered which enabled the natural rubber to be hardened, which solved the problem of its stickiness.

(*ii*) The invention of the pneumatic tyre and the advent of the motor vehicle created the demand for rubber.

(*c*) *Petroleum.* Oil has been known (as bitumen) since Biblical times but its potentialities as a fuel were not realised until about a century ago when certain discoveries revealed its potential:

(*i*) A use was found for it as a fuel, with the invention of the internal combustion engine.

(*ii*) The process of distillation was discovered by technologists, which enabled crude petroleum to be broken down into fractions.

(*iii*) The science of geology became sufficiently advanced to enable petroleum deposits to be located.

Resources of this kind may be described as latent resources. Only when man finds a use for them and is capable of using them do they have any real meaning or value for him.

4. Science and technology. Professor S. H. Beaver has stressed that the geographical environment embraces not only the physical and bio-geographical environments, which together form the natural environment, but the human environment, i.e. the social, economic, political and technological aspects.

What we may call the technological environment differs from the other man-made environments (social, economic and political) in the sense that it is an outgrowth from, and a transmuting force for, the other aspects of the human environment. As Beaver aptly writes: "The technological environment acts as a filter between man and all the other aspects of the total environment, controlling his ability to adapt himself to them and to modify them" ("Ships and Shipping: the Geographical Consequences of Technological Progress." *Geography*, April 1967).

Science and technology have been responsible for the change in the relationship between man and his environment. Early man, with his limited knowledge and his limited technical equipment, lived largely on the sufferance of nature and was subject to it; modern man, in contrast, attempts with considerable success to dominate and control nature. Science and technology may be regarded as the twin keys unlocking the door to human progress, prosperity and well-being.

In **5–16** below we shall briefly consider the earth's natural resources in turn.

5. Soil resources. Soil is a fundamental resource and of tremendous significance to man and to almost all living organisms since the bulk of the world's food is grown in the soil or comes from animals which feed on the vegetation growing in the soil. The soil is also the basic source of a wide range of industrial raw materials, e.g. timber, cotton, tobacco, vegetable oils.

Soils vary widely in their character and fertility; for example, podsols and most red and yellow tropical soils are leached and relatively infertile whereas chernozems and chestnut soils and most desert soils are rich in mineral matter and, given water, are highly productive.

Soil is a substance of much greater complexity than one realises. It is composed of inorganic matter (broken, pulverised rock) and organic matter (humus or decayed vegetable matter), together with air and water. In it live small animals and numerous bacteria which help to create soil and keep it in good condition. Soil is not dead, lifeless, immovable or inexhaustible; we should regard it rather as being dynamic and destructible. Soil is slow to form (on average about 1 cm every fifty years) but easily lost. Under certain conditions soil can be lost, a process known as soil erosion (*see* **6** below).

6. Soil erosion. Soil erosion occurs for two basic reasons:

   (*a*) When the soil is washed away by torrential rains.
   (*b*) When the soil is blown away by strong winds.

In many tropical regions where the forest or grass cover has been removed, the heavy rains have deeply gullied the land or removed the fertile topsoil by sheet-wash. In semi-arid and temperate grassland regions, overgrazing and ploughing the prairies has exposed the ground to wind attack and large areas, such as the "dust-bowl" area in the west of the United States, have been denuded of their topsoil. In view of the world's rapidly growing population, the world cannot afford to lose even 1 km^2 of soil, but more soil is being lost each year—through soil erosion and other wastage (such as building on good soil)—than nature makes; hence the soil is a shrinking asset.

Fortunately, man has learned how to combat soil erosion, although of course it does not always follow that he adopts practices which check erosion. There are still many farmers in the world who are content to "mine" the soil, to rob it of its fertility, and who do not, or cannot, care for their soil properly. Soil requires very careful use and management for its maintenance.

7. Vegetable resources. Cultivated crops apart, the following are the earth's principal vegetable resources:

(a) Natural grazing lands (see **8** below).

(b) Forests (see **9** below).

(c) Marine plants (see **10** below).

8. Grazing lands. Many animals, especially the world's beef cattle and wool sheep, depend upon the natural pastures of the prairies, pampas, steppes, savannas, etc. Many wild grass-eating animals (herbivores) also depend on nature's grazing grounds. The reindeer and other tundra animals are dependent upon the arctic pastures—the mosses, lichens, herbaceous plants, etc. The natural grasslands in themselves yield few commercial products —esparto grass is one of the most important—but they are essential for large-scale pastoral farming.

Many of the world's natural pastures are suffering from neglect and progressive deterioration. The chief problems arise from the following causes:

(a) The encroachment of the deserts along the margins of the grasslands.

(b) Overgrazing by stock which is gradually reducing the carrying capacity.

Man could do much to restore and improve these natural pastures: by limiting the numbers of stock to the carrying capacity of the land; by no longer ploughing up grassland in marginal areas; and by attempting a certain amount of judicious reseeding.

9. Forests. The world's forests supply man with two principal groups of commodities:

(*a*) Timbers, both hardwoods and softwoods of various kinds, used for constructional work, furniture, woodpulp and, sometimes, fuel.

(*b*) Forest products, such as vegetable oils and waxes, rubber, gums, resins, bark and fibres together with an assortment of drugs and medicines.

The world's forests yield many natural products which have uses as foods, medicines or industrial raw materials.

Forests formerly covered a much larger area than they do today, but the land was cleared because of the need of land for crop cultivation or increased pasture for animals and because of the demand for timber for constructional purposes, for fuel and charcoal-making and, more recently, for woodpulp for paper and rayon manufacture. The demand for timber today is greater than ever before, notwithstanding the growth of synthetic products and substitute materials. It is clear that a timber famine is likely in the future unless the wholesale depletion of forests is stopped and a programme of afforestation is vigorously adopted. Many countries, especially those that depend to a great extent upon their forest resources, such as Norway, Sweden and Finland, have adopted stringent forestry laws to conserve their timber resources; they also follow a system of re-planting.

10. Marine plants. Some specialised types of plants live in fresh or sea water. Minute plants provide some of the food for fish. A few water plants have some commercial value. Perhaps the most important is seaweed; considerable use is made of it for food, for medicines and as manure for crop cultivation; it is also used in the manufacture of soaps, jellies and paper varnish.

11. Animal resources. Man makes use of the animal world for a variety of reasons but the principal uses are threefold:

(*a*) Foodstuffs, e.g. fish, meat, milk, eggs, honey.
(*b*) Raw materials, e.g. wool, hair, hides, horn.
(*c*) Carrying and draughtage.

In earlier times, many peoples lived by herding animals, as a few peoples do even today, e.g. the Bedouin (by sheep, goats, camels), the Masai (by cattle) and the Lapps (by reindeer); but nomadic pastoralism as a way of life is rapidly declining. On the other hand, man is becoming increasingly dependent upon certain type of animals for foodstuffs and raw materials. Some species have become extinct, and others nearly so, through ruthless hunting.

Fish and other marine creatures are included among the animal resources and yield important supplies of fish-food, fish-oil, fertiliser and lesser commodities of a precious or semi-precious character, e.g. coral, pearls, ivory, fur.

12. Animal breeding. Man, as a result of long experience of breeding and, more recently, as a result of the practice of artificial insemination, along with the use of new types of feeding-stuffs, is able very largely to control the numbers and characteristics of his domesticated animals. For example, he is able to do the following:

(a) To crossbreed stock which can live more successfully in tropical environments.

(b) To breed sheep which yield improved fleeces.

(c) To breed cows which give greater quantities of, and richer, milk.

(d) To rear pigs that grow exceptionally fast on less food.

(e) To produce animals which yield more lean meat and less fat.

Animal husbandry, indeed, has been completely revolutionised in recent decades.

Against the above advantages of factory farming must be set certain disadvantages: the hazards created by the use of chemicals in an attempt to increase food output, and the production of animal strains by artificial breeding that may be unable to resist the normal diseases and ailments which affect domestic animals.

While man has paid much attention to his domestic animals he has, paradoxically, paid scant attention to wildlife and even wantonly destroyed much of it. Many useful and valuable species are becoming extinct owing to ruthless hunting or to the destruction of their natural habitats. Fortunately, man is now beginning to appreciate the importance of wildlife and steps have already been taken to help preserve some of the declining animal population. More, however, needs to be done and there is an urgent need for closer international control of world fishery resources.

An aspect, too, of which man is only just beginning to be aware is the fine state of balance which exists in nature. If man upsets this balance by killing off (through his use of chemical fertilisers, insecticides, etc.) insects, birds and other creatures, he may create even greater difficulties for himself in the future.

13. Minerals. A great variety of mineral resources of great use to man and which may be said to be indispensable in modern civilisation are found scattered throughout the earth's crust. These mineral resources may be divided into four main groups:

(*a*) *Mineral fuels*, such as coal and petroleum.

(*b*) *Metals*, such as iron, copper, tin and gold.

(*c*) *Non-metallic minerals*, such as salt, sulphur, and nitrate.

(*d*) *Constructional materials*, such as sandstone, granite, and clay.

The occurrence and distribution of minerals varies widely. Some, such as clay, gravel and building stones are widely spread and abundant except locally. Other metals, such as iron, are widely spread and relatively plentiful, although workable concentrations are of relatively restricted occurrence; others, on the other hand, such as cobalt and nickel, are of limited occurrence.

Some metals, such as iron and copper and lead, have long been used and in appreciable quantities; others, such as aluminium and tungsten, have been used only in the present century. The use and value of particular metals is closely linked with technological advances; for example, the development of electricity led to a great increase in the use of (and also in the value of) copper, while the electricity industry found a use for tungsten; tungsten also became of enhanced importance with the growth in the demand for special steels, especially steel used in the making of high-speed cuttings tools.

14. Mineral conservation. Minerals differ from most other resources in that they are essentially exhaustible and the earth's stock of mineral wealth is quickly becoming depleted. Already many deposits of certain minerals have been exhausted—the increasing international trade in iron ore is a result partly of local exhaustion of iron ores—and there is a certain anxiety about some mineral wealth, more especially the fossil fuels. Since minerals are irreplaceable, it is important that they should be used sensibly and not squandered, especially those that are in relatively limited supply.

The conservation of minerals can be undertaken in various ways:

(*a*) Care must be taken not to use them wastefully as has often happened in the past.

(*b*) Improved methods of extraction must be adopted wherever possible.

(*c*) The efficient recovery of minerals that have already been used must be practised.

(*d*) Substitute material should be used, wherever this is possible.

By adopting these methods great economies could be effected in the use of mineral wealth.

15. Water. Water is one of the most important and most precious of all the natural resources. Not only is it indispensable for life but it has numerous other uses, listed below:

(*a*) *Agriculture*—watering stock and crop growing, especially in arid and semi-arid regions where irrigation is required.

(*b*) *Industry*—for cleaning and processing commodities, for steam generation, for cooling purposes.

(*c*) *Power*—in earlier times running water turned waterwheels but nowadays the water is mostly used in the production of hydro-electric power.

(*d*) *Transport*—waterways (rivers or canals) are utilised for the cheap carriage of commodities in bulk.

(*e*) *Effluent disposal*—to transport wastes from urban and industrial areas.

(*f*) *Food supplies*—to provide habitats for fish which in some areas may be significant as an article in the diet.

(*g*) *Domestic uses*—drinking, cooking, bathing, washing, sanitation, garden watering.

(*h*) *Municipal uses*—fire fighting, street cleansing, in hospitals and schools, ornamental lakes and fountains.

(*i*) *Recreation*—water may have an amenity value, providing opportunities for sporting activities or merely serving an aesthetic purpose.

16. Water conservation. Water is not in any real sense short: there is plenty of it—even fresh water—in the world, but locally, or regionally, water supplies may be short. In countries like Britain, where there is normally plenty of rain, water is taken very much for granted, used liberally and not infrequently wasted. In contrast, in dry countries, such as the Middle East countries, water is in short supply and is a precious commodity. The overall world demand for water is, however, increasing rapidly and we are now reaching the stage where we are becoming water conscious and realising that there is a need for the careful regulation and conservation of water supplies. It must be recognised that the provision of extra supplies of water often incurs great capital cost for dams, reservoirs, treatment plant and pipes.

Water, particularly in areas where there is pressure upon supplies and where many varying demands are put upon the available resources, could be conserved by these means:

(*a*) Metering water supplies and increasing the cost of the water.

(*b*) Limiting pollution through controlling the amount of effluent discharged into rivers.

(*c*) Recycling water, particularly in those industries which use large quantities of water.

(*d*) Making more efficient use of the water that is used.

(*e*) Preventing waste through pipe leakages, evaporation and so on.

(*f*) Recharging sub-surface supplies of water to ensure continuing resources.

ECONOMIC ACTIVITIES

17. Basic human needs. Certain basic needs must be satisfied in order that man may live (*see* **1**). Most of man's time and energies are spent in supplying these fundamental needs. The ways and means he adopts to satisfy these wants are many and vary between place and place.

In some environments the physical conditions of relief, soil, climate and vegetation make life extremely difficult: nature limits the available resources by which man can live. On the other hand, nature offers much less restrictive conditions in other areas and man find it much easier to live; nature, being more bountiful, offers a wider range of resources and opportunities.

18. The use of the environment. The uses to which man puts his environment are the outcome of two influences:

(*a*) His social organisation.

(*b*) His cultural development.

The more man learns concerning the ways of using what his environment offers, the easier will his life become. If man merely lives off the land, i.e. merely collects the fruits of nature, his existence will be hazardous. Once he has learned the art of cultivation, however, his food supply becomes more assured. Food production, as distinct from food collecting, relieves him of the necessity to move around constantly from place to place; it enables him to live a sedentary life where he can live and work in one place.

But man uses his environment for things other than the provision of food, water, clothing and shelter. Today, with money to spare and increased leisure time, he uses his environment for sporting activities and holiday-making (*see* X). He also uses his environment for scientific experiment and the pursuit of knowledge.

19. The dawn of civilisation. The discovery of the art of crop growing was one of a number of developments (including the domestication of animals, the making of pottery and the weaving of cloth) which ushered in the New Stone Age or the Neolithic Revolution, somewhere about 7,000 or 8,000 years ago (*see* III). These developments, which seem to have occurred in the Near and Middle East, had very important effects upon human life and may be said to have formed the basis of civilised life, as distinct from the barbarism and savagery which had previously prevailed.

The art of cultivation made possible the following:

(*a*) Man could live in one place permanently.

(*b*) Large groups of people could live together.

A sedentary and communal life allowed other things to follow:

(*a*) Man could have a permanent dwelling and could acquire possessions.

(*b*) Specialisation of labour, i.e. division of work, became possible.

(*c*) Division of labour brought increased leisure which allowed artistic and literary developments to take place.

From this cultural revolution civilised life developed and a whole series of civilisations emerged, e.g. the Egyptian, Greek and Roman. Western civilisation as we understand it today is based upon, and has grown out of, these early cultures.

20. Cultural paradox. During several millennia of development, man has progressed by discovering and inventing. The twentieth century, with its electronic devices, supersonic jets, nuclear power, etc., is a far cry from the Neolithic Age. And yet, even at the present day, there are peoples in various parts of the world who live in much the same way as the peoples of the Neolithic

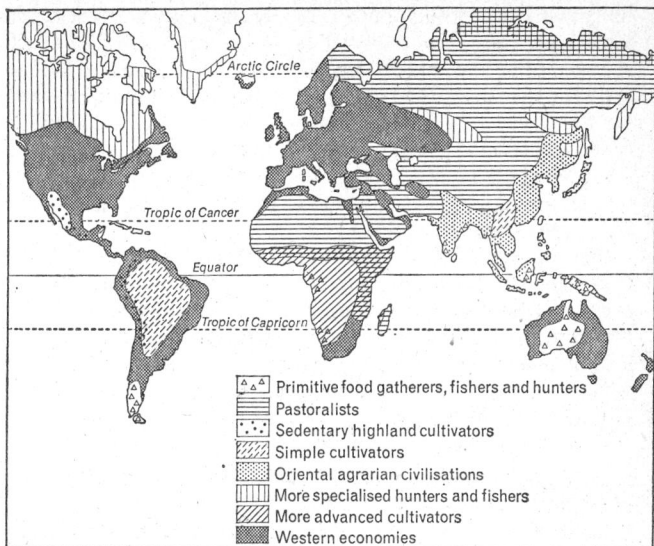

FIG. 12 *The principal economies of mankind.*

This map, showing the principal modes of life, is of course generalised. Much of Asia is shown as being under pastoralism but there are many localities where cultivation takes place, as in the oasis settlements of central Asia. Moreover, the map takes no account of mining or industrial activities which clearly are widespread in Europe and the United States and elsewhere in the world.

period. Here is one of the great paradoxes of our time that side by side with the scientific and technologically advanced societies there are other communities who base their existence upon the simple economies which prevailed in Neolithic times.

Let us now look at the different ways in which man, both in the past and at present, has lived and earned his livelihood. The world distribution of man's principal economies is shown very generally in Figure 12.

SIMPLE SOCIETIES

21. Food gatherers and hunters. In the very earliest stages of his cultural development, man was a food collector and petty hunter;

he gathered wild fruits, nuts, roots, grubs and wild honey, and undertook fishing, snaring small fry and hunting. Living by such simple means, and entirely dependent upon the bounty of nature, it meant that man was compelled to live in small groups—perhaps limited to the family group—simply because the available food supplies in any area were incapable of supporting more than a few people. The size of the group, or in other words the density of the population, was limited by the productivity per square kilometre of land.

Small groups of people continue to follow this mode of life but they usually live in the more remote and inaccessible areas. However, as civilisation spreads, less and less territory is available for the relic groups and they are being driven into the more inaccessible and less hospitable areas. How long such social groups will continue to exist is problematical but it would seem unlikely that it will be much longer.

Certain examples are as follows:

(a) Some of the Eskimo and Indian tribes of the arctic regions of North America.

(b) The Yaghans of Tierra del Fuego in the cold, bleak southern extremity of South America.

(c) The pygmies of the hot, wet forests of the Zaïre Basin and south-east Asia.

(d) The primitive but hardy Bushmen of the Kalahari Desert in south-west Africa.

(e) The aborigines of Australia who dwell in the northern and semi-arid lands of that continent.

22. Primitive cultivators. Many people are simple farmers, growing crops upon which they subsist. Crop growing is a stage more advanced than food gathering, but the primitive cultivator may be a food collector and hunter also. Sometimes even herders grow crops if conditions are suitable and they sojourn a while in a given spot. This shows the inadequacy of trying to classify human activities too rigidly; many people, especially at the primitive level, are seldom exclusively food gatherers, fishers, hunters, herders or cultivators.

Simple cultivation is of either the shifting or the sedentary type (*see* **23** and **24** below).

23. Shifting cultivation. This, widely practised in the tropical zone, involves the clearing of the vegetation cover, usually by the "slash and burn" technique, the planting of seeds, tubers or cuttings either by crude hoeing or simply by making holes in the ground with a digging-stick and subsequently the harvesting of the crop. Such primitive tillage knows little or nothing of the plough, the rotation of crops, or the use of fertilisers. After two or three years of cropping, the soil becomes exhausted and the cultivator clears a fresh patch of land. This practice is sometimes called milpa cultivation. Often the clearing of the land is done by the menfolk, while the women plant and tend the crops. Examples of peoples living by shifting agriculture are these:

(a) The Indian tribes, such as the Boro, of the Amazon Basin who grow manioc, sweet potatoes, yams and beans.

(b) Some of the Negro peoples of west and central Africa, such as the Yoruba of Nigeria and the Boloki of Zaïre, who grow yams, maize, millet and bananas.

(c) Various hill tribes of south-east Asia, such as the Liao, Moi and Lolo of Laos and the Muruts of north Borneo, who grow rice and beans.

24. Sedentary cultivation. Settled cultivation is usually of a more developed kind. Sedentary agriculture, however, varies widely, ranging from the backward methods pursued by many tropical cultivators to the skilled, highly intensive tillage practised by many oriental subsistence farmers. Peasant agriculturalists often have to work with simple, sometimes crude, implements although some kind of plough is normally used, even if it is a hand plough. Although manures (animal and green) may be used, chemical fertilisers are seldom used, if only for the simple reason that the peasant cultivator cannot afford them. A variety of crops is grown, such as rice, maize, millet, ground-nuts, beans and other vegetables.

Rarely is the sedentary cultivator purely a subsistence farmer; very often he grows a small crop, perhaps cotton or oil palm, as a cash crop. The peasant farmers of Latin America, Africa and Monsoon Asia are commonly sedentary cultivators. The most highly developed form of sedentary agriculture occurs among the peasant farmers of eastern Asia. The Chinese, Japanese and some of the peoples of south-east Asia developed a highly skilled and very productive agriculture using such techniques as inter-cropping, crop rotation, fertilisation and irrigation. Returns are high but the farming is labour intensive; in other words, output per unit of land is high but output per capita is low. This is largely explained by the unmechanised character of the agriculture.

Traditional sedentary cultivation is, however, being transformed in many places. In China, for, example, traditional peasant farming no longer exists since the Communists expropriated all the land, set up collectives and finally (by 1963) centralised communes. Of the 74,000 communes established, each, on average, consisted of some 30 co-operatives involving some 5,000 households and some 836 ha of agricultural land. Unified systems of cultivation were imposed on the communes. The "Green Revolution" in India has also had a considerable impact on farming: as a result of the introduction of high-yielding varieties of rice and wheat, together with a rapid increase in the use of chemical fertilisers and pesticides, and the expansion of irrigation, food grain production has increased substantially. The most notable advances have been made by the bigger farmers possessing some capital which has enabled them to apply the new methods.

25. Pastoralists. The domestication of animals meant that man, instead of hunting animals, began to herd them. This involves the control and protection of animals in order that man may live off them through the milk, meat, wool, hides, etc., which they provide. Some wild animals, e.g. cattle, sheep, goats, horses and camels, were amenable to domestication and these man tamed and herded. Such creatures were especially useful to man for they provided the following:

(a) Food (milk, meat and blood).

(b) Raw materials (hides, hair, wool, horn).

(c) Carriage and draughtage.

Such animals were herbivorous, i.e. grass-eating, and they wandered over the pasture lands in search of forage. In the beginning, man wandered with them, going wherever the animals went; later he organised or systematised their movements so that definite seasonal migrations from area to area took place. Such wandering movement is called nomadism (*see* **26** below).

We may take as an example of pastoralists the Fulani and Masai of the Sudan. These Negro tribes rear herds of cattle or, in the drier areas, herds of goats. They count their wealth in the number of animals they possess. The Masai frequently drink the blood of their cattle. The Masai tend to be sedentary rather than nomadic herders.

26. Nomadism. Nomadic herders keep to their own well-defined territories and follow a fixed and predetermined pattern of movement. To wander at random into another tribe's preserve is to court serious trouble. The seasonal migrations may involve treks of hundreds of miles.

Formerly most pastoral people were nomadic; a few groups remain so still, though the numbers engaged in nomadic herding are declining rapidly.

(*a*) The Bedouin of the Middle East continue to wander with their camels and horses from oasis to oasis. They breed camels and horses which they sell to the settled peoples living around the desert margins. In the past they also carried trade and were not averse to occasional raiding. They live largely on milk and dates together with grain bought in the oases.

(*b*) The Kirgiz of the Central Asian steppes were until quite recently nomads rearing horses and sheep, but their migratory mode of life has been changed during the past generation or so by the Soviet Government which has "collectivised" animal herding. Thus the Kirgiz are now sedentary pastoralists.

(*c*) The Lapps, Samoyeds and Tungus, who live on the northern tundra margins of Eurasia, are rearers of reindeer. The reindeer provide milk, meat, hair and hide and are the basis of this nomadic existence. In summer the animals feed on the mosses and lichens of the tundra pastures; in winter they seek the shelter of the forests.

ADVANCED SOCIETIES

27. Commercial activities. Most people live in societies which are more developed and more complex than those described above; many are still concerned with food production but a large number earn their livelihood by extractive activities, by manufacturing industries, or by performing services of some kind. The economies of advanced societies are very largely based upon exchange; they produce primary products or manufactured goods for sale. There are some countries, such as Uruguay and New Zealand, which are preponderantly concerned with the production of primary produce and import many of the manufactured goods they need. Conversely, there are some countries, such as Britain and Belgium, which are essentially industrial countries and which are dependent to a considerable extent upon the importation of foodstuffs and primary products.

28. Commercial fishing. The production of fish and other marine products for sale differs from primitive fishing, which is primarily for subsistence, in both its scale and organisation. But, in one sense, the commercial fishers have advanced no further than the primitive fishers for they still hunt rather than rear fish. There are cases of man, breeding and rearing fish—in fact, the Chinese have been fish-farming for some two thousand years—but this is still relatively small-scale and localised. However, we are reaching the stage where man will soon have to manage the world fisheries on a more scientific basis.

Fishing on a commercial basis is normally a highly organised activity employing specialised craft and techniques. The chief areas fished are the shallow seas on continental shelves in the temperate waters of the northern hemisphere, although there are other productive grounds in the southern hemisphere which, so far, have been little exploited. Some of the more important fisheries are these:

(a) The North Sea, formerly a rich fishing area, now has depleted resources through over-fishing, and herring fishing has had to be temporarily banned. Norway, Denmark and Britain, however, still have substantial fishing industries.

(b) The waters around Iceland are rich in fish and Iceland is greatly dependent upon its catches.

(c) The Newfoundland—Nova Scotia area which has renowned cod fisheries, although there is inshore fishing for molluscs and crustacea.

(d) Off the coast of British Columbia and southern Alaska there are the salmon and halibut fisheries.

(e) The fisheries of the Japanese archipelago which yield some 10 million tonnes of fish each year. Japan now leads the world as a fishing country.

(f) The Peruvian fishery, which until a few years ago had grown to be the largest in the world, has substantially declined on account of the shrinkage of the anchovy shoals. Production has dropped to half, i.e. 4 million tonnes; however, the catch is mainly processed for fish meal (used as a fertiliser).

Whaling, sealing, sponge collecting and pearling are commonly looked upon as branches of the fishing industry. All these, however, are declining activities. The fishing industry has given rise to many ancillary activities, e.g. barrel-making, canning, fish-oil extraction, the making of fertilisers and the manufacture of fishing nets.

29. Commercial grazing. Large-scale commercial grazing is the chief form of land use in the world's drier areas where tillage and forestry are not practicable, i.e. in continental interiors where the rainfall is light and occurs seasonally.

(*a*) *Limited nature.* Commercial grazing is practically confined to cattle and sheep rearing, the two domesticated animals of greatest value for the production of meat, dairy produce, wool and hides, primarily for export purposes. The industry has been helped by a series of inventions and developments which include the following:

(*i*) The introduction of wire fencing which enabled the ranges to be enclosed.

(*ii*) The use of alfalfa and other plants which have improved feed.

(*iii*) The introduction of pesticides which have helped to reduce the ravages of the cattle tick and other pests.

(*iv*) The development of refrigeration processes which enabled meat and dairy produce to be shipped over long distances.

(*v*) The introduction of the tin can which made possible the canning of meat.

(*b*) *Recent development.* Commercial grazing is an economic activity which has developed largely over the past hundred years. It has been closely related to these developments:

(*i*) The industrialisation of western Europe in particular.

(*ii*) The growth in population.

(*iii*) Higher living standards which have created the market.

(*iv*) A series of scientific and technical inventions which have enabled animal products to reach distant markets in good condition.

(*c*) *Commercial grazing.* This is carried on in these areas:

(*i*) *The drier temperate grassland areas*, e.g. the Great Plains of North America, the pampas of Argentina–Uruguay, the veld of South Africa and the interior lowlands of Australia.

(*ii*) *The tropical grassland or savanna areas* of some tropical countries, e.g. the campos of Brazil, the llanos of Venezuela and in Queensland in Australia, although the industry is not as well-developed in these areas.

30. Commercial agriculture. This differs from subsistence farming in several respects:

(*a*) All, or a substantial proportion, of the produce is for sale, either in the home market or abroad.

(*b*) Scientific methods, such as crop rotation and the use of insecticides, are consciously applied.

(*c*) Fertilisers, either natural or artificial, are commonly used in substantial quantities.

(*d*) Farming practice is mechanised to a greater or lesser extent.

(*e*) Special strains and improved varieties of both plants and animals are raised.

As a human activity commercial farming is becoming increasingly specialised, e.g. market gardening, fruit-growing, intensive dairying. Specialisation in one form is the concentration upon a particular crop, as in the case of plantation products. Such single-crop cultivation is termed monoculture. Monoculture is often a highly lucrative form of farming, although it is rather precarious, for a fall in the world market price or an attack by a plant disease may bring ruin; for example, the collapse in the price of natural rubber in the early 1930s brought great difficulties to the producers in south-east Asia, while in the Caribbean the ravages of the sigatoka disease destroyed the banana plantations. At the other extreme is mixed farming, which is often practised in Britain, where not only is there a variety of crops grown but also a variety of animals may be raised. Such a system gives greater economic security, although the returns may not be as lucrative.

The application of science and technology to agriculture is revolutionising the industry and reducing drastically, in many instances, the need for manpower. Britain today, for instance, is producing more food and farm products than ever before, yet her agricultural work-force is less than a quarter of what it was at the beginning of the century.

31. Influences on commercial farming. Commercial farming is influenced by four factors, chiefly physical, biotic, human and economic.

(*a*) The main physical conditions are as follows:

(*i*) The topography, i.e. the height, ruggedness.

(*ii*) The soils, i.e. the fertility, workability.

(*iii*) The climate, especially the length of the growing season.

(*iv*) The water supplies, provided by either rainfall or irrigation.

(*b*) Biotic factors are of great importance, for animal and insect pests may handicap, even preclude, particular crops, or some kinds of livestock from being reared; e.g. the boll weevil has attacked the cotton plant, the tsetse fly prohibits cattle rearing.

(c) Human factors include the following:

(i) Experience.

(ii) Tradition.

(iii) Skill.

(iv) Energy.

(d) Economic influences ultimately determine whether or not farming will be carried on:

(i) Production costs must leave a margin of profit.

(ii) Transportation facilities must be available to dispose of the produce.

(iii) There must be a demand for a particular commodity to provide a market.

32. Forestry. Forestry is a distinctive extractive industry mainly associated with advanced societies. In most areas where there is lumbering the activity is highly mechanised. Machines cut, saw and transport the timber. Only in tropical lands are more primitive methods used. The industry, too, is becoming more systematic and scientific, a necessity forced upon it by the continuing, and indeed increasing, demand for wood, and the need to replenish the forest resources by afforestation.

Usually forestry is a seasonal occupation, largely because many of the forests lie in high latitudes and there are climatic problems. In Scandinavia–Finland, the Soviet Union and Canada exploitation is largely seasonal. In some cases forestry is associated with agriculture, as occasionally happens in Britain.

The principal exploitable forests are as follows:

(a) *The coniferous softwood forests* of Canada and northern Eurasia which yield constructional timber, wood-pulp, tar and other forest products.

(b) *The temperate hardwood forests*, now much depleted, of temperate latitudes, providing certain timbers for furniture making, e.g. oak, walnut, together with bark and nuts.

(c) *The monsoon deciduous forests*, chiefly of south-east Asia, yielding especially teak, a valuable hardwood, but also bamboo and rattan.

(d) *The equatorial forests* of Amazonia, west and central Africa, and the east Indian region, yielding prized cabinet woods, wild rubber, nuts and medicinal plants.

The softwood forests of the northern hemisphere are the ones which are exploited most since they are nearest to the great consuming centres. They are important not only for their timber but also because they are providers of wood-pulp which is in great demand for paper-making and rayon manufacture. In the southern hemisphere there are smaller, but valuable, forest areas as yet little exploited, partly because they lie too distant from world markets. The stands of Parana pine in southern Brazil, however, and the jarrah and karri forests of Australia are exploited while the temperate forests of southern Chile are almost untouched.

33. Mining. Mining is an activity which has taken place for thousands of years, from the time when man first learned the value of metal for making implements. The great development of this extractive industry came with the Industrial Revolution and the needs of our modern machine age. It becomes increasingly important as greater quantities of minerals are used. Many people, though their total numbers are comparatively small, earn a livelihood by mining. Four important points with respect to the mining industry should be noted:

(a) Mineral resources are widely scattered and often highly localised; therefore mining is a form of economic production which is often mixed with cropping, grazing and forestry. Because of the localised occurrence of minerals, there are often concentrations of population in areas which otherwise are unpopulated or very scantily peopled.

(b) Mineral wealth is the basis of much modern industry; hence considerable numbers are engaged in mining either for fuels, such as coal and petroleum, for metals, such as iron and copper, for non-metallic minerals, such as salt and phosphate, or for building materials, such as sandstone and clay.

(c) The mineral fields have often become magnets for industry and, therefore, great concentrations of population. This is more especially the case with the coalfields, e.g. the Ruhr and Pennsylvania coalfields, which during the past hundred years or so became *the* great foci of manufacturing activity.

(d) Mineral wealth is exhaustible; hence mining is a "robber economy". Once used the resource is exhausted and, unlike animate resources, cannot be replenished.

Mineral wealth and mining show interesting geographical responses and relationships. In early times man travelled far in search of tin, essential in the making of bronze. Precious metals, particularly gold, have throughout history lured man to foreign parts, e.g. the Spaniards to Mexico and Peru. Very valuable minerals of small bulk, such as diamonds and gold, can be transported far since freight costs in relation to their value are low. In contrast, coal and iron ore are bulky, low-value commodities and, although some is transported over long distances, usually they are used on or near the site of extraction, e.g. the smelting of the low-grade Jurassic iron ores at Corby.

The value and importance of specific minerals has led man into the most inhospitable of human environments: think of oil mining in the Sahara, of uranium in the Canadian barren lands, and of tin extraction at altitudes of 4,750 m in the Andes. Many large cities owe their existence to mineral wealth, e.g. Johannesburg to gold, Dallas to oil; on the other hand, the exhaustion of mines has produced "ghost towns" in some areas. (*See* also **13**.)

34. Manufacture. In contrast to all the foregoing types of activities and forms of production, manufacture is a secondary activity since it uses the primary raw materials from the sea, the land, field and forest, processes them and converts them into commodities more useful to man.

In the widest sense manufacturing has a world-wide occurrence. Even the most primitive communities engage in some form of manufacture, e.g. the making of tools, pots and cloth. But such "domestic" or "cottage" industries stand contrasted with the highly complex manufacturing activities of modern communities.

More people are employed in manufacturing than in any other activity, save agriculture. In the more advanced countries, manufacturing is the whole-time occupation of the greater proportion of the population. This is true even of such countries as Denmark and the Netherlands which one tends to regard as being more particularly farming countries.

Although modern manufacturing industry, and more particularly what we have been prone to call "heavy industry", developed mainly on the coalfields, since they provided the chief source of energy, there has been in more recent times a distinctive move towards the decentralisation of industry. This wider spread of manufacturing industry has been due to several factors, including economic and political, but mainly it has resulted from two developments:

(*a*) The greater flexibility in transport provided by motor transport.

(*b*) The use of electric power which can be carried widely by the grid system.

It has been said that the development of nuclear power will enable industry to become footloose. Certainly such power is likely to become more important in the future and will make industry less tied to traditional sites, but it would also seem unlikely that the present-day great centres of industry will be replaced because of all the other accrued advantages which they possess.

35. Commerce. While the majority of mankind is engaged in primary production of one kind or another or in the secondary activity of manufacturing, a large and increasing number of people, especially in the more advanced countries, are concerned with activities which do not add directly to the total volume of goods produced. Such people as, for instance, those engaged in buying and selling, in banking, financing and insurance, in transport and commerce, are equally as necessary as "producers" in modern society.

Commercial activities, those organising, facilitating and transacting exchange, fall into two groups:

(*a*) *Traders*, i.e. those who act as middlemen between the producer and the consumer, may be said to include, in the broad sense, merchants, shippers, brokers, financiers and bankers.

(*b*) *Transporters*, i.e. those who engineer the movement of goods and people from place to place, e.g. lorry drivers, pilots, merchant seamen and those concerned with communications, e.g. radio operators, telephonists.

Activities of this kind are usually termed tertiary activities.

There is also a large number of people, especially in advanced societies, employed in the professions or servicing occupations. These people are not producers in the strict sense of the word. To this group of human activities belong scientists, doctors, lawyers, teachers, ministers of religion, administrators, policemen, artists, musicians, writers, entertainers and many others.

THE UNDERDEVELOPED COUNTRIES

36. The meaning of "underdevelopment". Most people these days are aware, to a greater or lesser extent, of a simple, but undeniable, economic fact: that some countries are economically and technically well developed and their peoples enjoy high standards of living while, on the other hand, there are other countries that are underdeveloped and backward by twentieth-century standards and whose peoples are poverty-stricken, often illiterate, and suffer low living standards. It has become customary to refer to these two groups as the "haves" and the "have-nots".

There is, however, some difficulty in interpreting the term "underdevelopment" or, as it is applied to countries, "underdeveloped". In one sense underdevelopment may be taken to mean unrealised potential, but perhaps the best and most intelligible interpretation may be taken as the condition in which per capita real incomes in a country are low—low compared with the per capita real incomes in the "advanced" or "developed" countries such as the United States or most of the countries of western Europe. Defined in this sense, underdeveloped is synonymous with "poor", and national poverty may be taken as best interpreting the condition of underdevelopment.

NOTE: (*i*) Some people advocate the use of the term "developing" as being more appropriate than "underdeveloped". This, it is true, is more acceptable to countries that are rather sensitive about being dubbed "underdeveloped"; but if we mean, as indeed we do, that the resources (human as well as natural) of these countries have been under- or inadequately developed, then the term "underdeveloped" is more accurate. (*ii*) Some people have suggested that the term "pre-industrialised" might be more appropriate and less offensive as a description of the condition of underdevelopment; but this is scarcely, if any, better, for it suggests that industrialisation is tantamount to economic development and a necessary condition for development and this cannot be accepted.

37. Living standards. In Britain living standards are tended to be measured in terms of motor cars, automatic washing machines, colour television sets, etc., and if people possess these material goods we say they enjoy a high standard of living. These things are *one* aspect of a living standard but the condition that really constitutes a high standard of living is one that ensures a sufficient and well-balanced diet, good housing, security of employment, and adequate social services (i.e. education, hospitals, water supplies, sanitation, old-age pensions). The "have-not" countries are the countries which *have not* these essential things

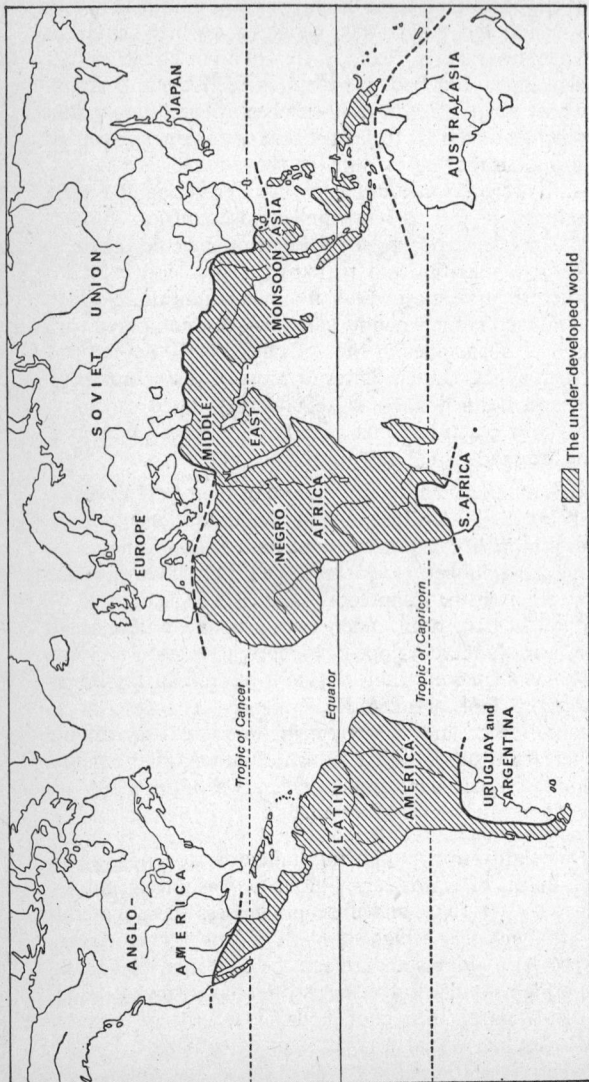

FIG. 13 *The developed and underdeveloped regions of the world.*

Note that practically all the developed regions of the world are those settled by white, or European, peoples. The Latin American region, Negro Africa, most of the Middle East and most of Monsoon Asia fall into the category of underdeveloped regions. There are, however, considerable variations in the under-developed world and some countries are now developing quite rapidly.

////// The under-developed world

that constitute high standards of living. Figure 13 shows the developed and underdeveloped regions of the world.

38. Characteristics of underdevelopment. A correlation exists between national poverty and many other features of the economic and social organisation of a country. Some of the more important characteristics of underdeveloped countries are as follows:

(a) A very high percentage (usually over 70 per cent) of the population is engaged in agriculture.

(b) The primary industries of farming, forestry, fishing and mining dominate the economy.

(c) An agricultural system employs large numbers of superfluous workers, i.e. a smaller labour force could produce the same total output.

(d) Farm holdings are small in size, agricultural techniques are primitive and the crop yields per hectare are generally low.

(e) Incomes per head of the population are low and there is very little capital per head.

(f) Birth rates and death rates are both high, the expectation of life is short and the population is frequently greater than the country can adequately support.

(g) There is a relatively low output of protein foods and consequently the people suffer from dietary deficiencies.

(h) There is overcrowding, bad housing and few public health services such as tapped water, proper sanitation, hospital services.

(i) Education services are poorly developed, usually at all levels, and there is a high degree of illiteracy.

(i) Women usually hold an inferior position in society and are not only denied equality but are frequently regarded as little more than chattels.

39. The underdeveloped countries. If we think of the underdeveloped countries as those which are poor and are experiencing acute difficulties in raising their standards of living, then we can say that most of the countries of Africa, southern and south-east Asia and Latin America fall into this category. Most, though not all, are located in the tropical belts and many of them were formerly colonies of the Western Powers. This distribution is shown in Figure 13.

It should be noted that the underdeveloped countries account for over two-thirds of the world's total population. A measure of their backwardness is that between them they have only 10 per cent of the total world energy consumption and account for a mere 7 per cent of the world's total manufactured goods.

There is, of course, a wide variation in the degree of underdevelopment: some countries, such as Brazil, are undergoing quite rapid development and the core area of Brazil, i.e. the hinterland of Rio de Janeiro, could now claim to be developed, as also could parts of Mexico and Iran. Other countries, however, such as the Sahel countries of Africa, Somalia, Bolivia and Paraguay, are making only slow progress.

THE PROBLEMS OF AGRICULTURE AND INDUSTRY

40. Agricultural problems. Most of the underdeveloped countries are basically dependent upon agriculture but the farming practices are backward and the output low in relation to the number of workers involved. A number of common handicaps, difficulties and problems occur:

(a) *Soils*, especially in the humid tropics, are poor, leached and soon become exhausted; hence yields tend to be low. Moreover, in areas of heavy, torrential rainfall soil erosion is apt to occur and indeed large areas have suffered on this account.

(b) *Pests and diseases* are prone to attack plants and animals in all tropical lands and so create serious problems for the agriculturalist. For example, sigatoka disease completely destroyed the banana plantations on the Caribbean coastlands while the tsetse fly prevents cattle raising over large tracts of tropical Africa.

(c) *Primitive techniques* of cropping and grazing (e.g. hoe cultivation, the absence of the plough, bush firing) result in low crop yields, poor quality animals, the spread of pests and diseases, etc. There is a general lack of scientific farm management, e.g. the use of fertilisers and mechanical aids to cultivation.

(d) *Social customs*, such as the division of holdings, the mortgaging of land to moneylenders, religious taboos, the prestige value of mere numbers of animals, the prevalence of the large estate and absentee landowners (especially in Latin America), often militate against efficient farming.

(e) *Subsistence agriculture* is characteristic over wide areas. This so-called self-sufficient type of farming is frequently self-*in*sufficient, for the cultivator often does not produce enough even adequately to maintain himself and his family, let alone produce a surplus for sale.

(*f*) *Crop specialisation*, or the dependence upon a single commodity, has been typical of many tropical countries, e.g. Egypt upon cotton, Sri Lanka upon tea, Malaysia upon rubber, Cuba upon sugar. This is too risky, for a fall in the world market price or a crop failure resulting from climatic hazards or pestilence and disease may bring disaster.

(*g*) *Transport and markets*. Even where a farmer is able to produce a surplus of a crop he often finds great difficulty in marketing it, since there may be a lack of facilities to transport it to the market or an absence of any marketing organisation.

41. Agricultural needs. The agricultural problem of the underdeveloped countries are manifold and complex. Changes in farming are likely to be slow, largely because farmers as a class tend to be conservative, but farmers will change their methods if it can be shown that they will benefit from doing so. It is important, however, that innovations should be geared to the local conditions: far too often in the past it has been assumed that the mere introduction of western ideas, techniques and management would bring an agricultural miracle.

A number of changes in the present agricultural system would seem desirable, however:

(*a*) *Agricultural education* seems to be essential if farming is to become more efficient and productive; this means more agricultural advisers, schools, colleges, experimental farms—all of which, however, cost money.

(*b*) *Diversification of agriculture*, especially in those areas where there has been undue dependence upon a single staple; but it must be recognised that this is a fairly long-term business.

(*c*) *Improved seed and animal strains* would help enormously to increase agricultural output, but for these the underdeveloped countries are largely dependent upon the agriculturally developed countries.

(*d*) *Fertilisers and insecticides* are required in large quantities to help fortify the inherently poor soils of the tropics and maintain soil fertility and to check the ravages of insect pests.

(*e*) *Agricultural machinery* could be of great benefit in many areas, enabling the peasant to cultivate more land and work it more expeditiously as well as relieving him of much of the toil and drudgery of farming.

(*f*) *Land tenure systems* in many areas need changing. The large estate, fragmentation of plots, indebtedness, all tend to hamper efficient farming, land care and improvement. Co-operative organisation would help in many areas.

42. Industrialisation. Almost all the underdeveloped countries wish to industrialise themselves: this is largely because they think industrialisation is a panacea for underdevelopment.

Some countries are content to industrialise gradually and pay due regard to their resources of power, mineral wealth and technological limitations; others are forging ahead rapidly, determined to industrialise no matter what the cost. Thailand and China represent these two extremes.

However, apart from one or two countries, such as India, Brazil and Egypt, which have achieved a considerable measure of success, most of them still have only very modest light industries, i.e. consumer goods.

43. Reasons for industrialisation. A variety of reasons help to explain this swing towards industrialism:

(*a*) It is recognised that no country can be powerful or politically secure without a well-developed industry, for industrialisation is a primary determinant of national power.

(*b*) The production mainly of primary products savours too much of the "colonial" era and psychological motives of this kind have helped to further industrialisation.

(*c*) Industrialisation offers one avenue of absorbing excess population, a problem which faces many underdeveloped countries, especially in Asia.

(*d*) Industrialisation is seen as a means of raising living standards.

(*e*) Industrialisation is a means of diversifying grossly unbalanced economies and assisting national aims of greater self-sufficiency.

44. Problems of industrialisation. Successful industrialisation implies not only the possession of adequate power, mineral and raw material resources but the acceptance of new values, an interested and mobile society, the availability of capital, industrial expertise, technological skill and an educated labour force. Without these, which usually necessitate radical changes in the economic, social and institutional structures of a country, industrialisation is likely to create more problems than it solves.

Some of the difficulties facing the underdeveloped countries in their attempts to industrialise are as follows:

(a) *Shortage of adequate power resources*. Since it is power potential above everything else which is a yardstick of a country's capacity for manufacture, it is essential that there should be adequate power resources available. Both Argentina and Brazil have made determined efforts to industrialise but time and again development has been handicapped by power shortages. Both have very little coal and it is of a poor quality. Brazil has a shortage of oil and must import the bulk of her requirements; Argentina, with reasonable oil and natural gas resources, is more fortunate. Brazil has a vast hydro-electric power potential, but the demand for electric power has outrun the country's capacity to produce it.

(b) *Shortage of capital resources*. Not only the development of power resources and the exploitation of natural resources but the setting up of industrial plants, especially for heavy industry, require huge capital expenditure. Many of the underdeveloped countries simply have not the capital to establish basic enterprises and are dependent upon financial aid from the developed countries. Most of India's iron and steel plants, for instance, have been established as a result of British, German and Russian financial and technical aid. Great schemes such as the Aswan High Dam, and the Volta Project, have only been made possible through the help of international aid.

(c) *Education and training*. One of the greatest difficulties facing the underdeveloped countries is their shortage of trained manpower. They suffer acutely from a shortage of technologists and technicians and have often to rely upon foreigners to manage their enterprises. Again, the industrial worker is frequently capable of doing only the simplest repetitive job and is unreliable. India's textile industry, for example, though long-established, has always suffered from a great turnover in its personnel. Without a consistent and reliable work-force, manufacture is bound to be inefficient. Not only is there a need for education and training at all levels but also for a loyal and stable work-force.

(d) *Political instability*. Political instability, which seems to be a malaise in many of the newly independent and underdeveloped countries, is a great handicap to economic development. The Latin American countries have long had a reputation for chronic political instability and to no small extent this has helped to keep them backward. The situation here, however, has tended to improve in recent times. Foreign investors are unwilling, naturally, to risk sinking their money in enterprises in countries whose futures are uncertain or whose rulers show no scruples. Internal unrest, constant changes in government, rioting, threats of expropriation and the like are not conducive to either investment or economic growth.

45. The need for industrialisation. Notwithstanding the many difficulties and problems besetting industrialisation in the underdeveloped countries, it seems fairly clear that most of them would benefit from some measure of it. It would, in fact, be economic sense to assist them in achieving this goal. Such help, granted, is likely to create new trade rivals in an already keenly competitive world, but it would at the same time lead to the development of new markets through the demands by the new wage-earners for consumer goods and by the states themselves for capital goods. Thus western standards of living would not necessarily suffer from the increasing industrialisation of the underprivileged countries. The strongest argument, however, for industrialisation is that it would help to promote better standards of living in countries where living standards are deplorably low. Higher standards of living would be likely to affect the social aspirations of the people, which, as past experience shows, help to check population growth.

To summarise, it would be desirable for the underdeveloped countries to adopt a measure of industrialisation where this is a feasible proposition: but these countries need the help, experience and financial assistance of the West to guide them in their efforts to industrialise, to promote their economic development, and to raise their living standards.

PROGRESS TEST 6

1. Explain the term "natural resource". What are the two main categories of resources? **(1, 2)**

2. Explain the importance of science and technology in man's conquest of his environment. **(3, 4)**

3. Explain the meaning of the following terms: effluent, latent resources, monoculture, milpa cultivation. **(3, 15, 23, 30)**

4. What is soil and why is it of such importance to man? **(5)**

5. Why are many of the world's pasture lands suffering pro-

gressive deterioration and what can man do to improve and restore them? **(8)**

6. How do mineral resources differ from soil, vegetable and animal resources? What can man do to conserve mineral resources? **(14)**

7. Name six ways in which water is used for the benefit of man. How can water resources be conserved? **(15, 16)**

8. What effects did the art of cultivation have on human life? **(19)**

9. Give some examples of primitive, simple societies and say where they are to be found in the world. **(21–26)**

10. Clearly differentiate between shifting agriculture and sedentary agriculture. **(23, 24)**

11. Locate the chief commercial fishing grounds in the world. **(28).**

12. Which inventions and developments have assisted commercial grazing? **(29)**

13. What factors influence commercial agriculture? **(30)**

14. Locate the chief areas of commercial forestry and name six forest products besides timber. **(32)**

15. Industry is no longer tied to the coalfields: why is this? **(34)**

16. Name the more important characteristics of underdevelopment. **(38)**

17. Which factors handicap farming in the underdeveloped countries? **(40)**

18. Why do backward countries try to industrialise themselves and what advantages does industry bring to them? **(42, 43)**

CHAPTER VII

Rural Settlement

TYPES OF SETTLEMENT

1. Settled life. In the early days of man's existence, when he was a collector and hunter of food or a nomadic herder, he seldom had a permanent home unless it happened to be a cave which he may, perhaps, have occupied from time to time. But under the simple economies of those far-off days there was little chance of man making a permanent settlement; such could hardly come until he had discovered ways of producing food from the land, i.e. the art of crop cultivation. Once he had learned how to till the soil and plant and grow crops, it became possible for him to have a permanent home and for him to create permanent settlements. This great change in human life, as we saw in III and VI, is known as the Neolithic Revolution which occurred somewhere around 5000 or 6000 B.C.

2. Permanency of settlement. Permanent settlement is related to two important conditions or considerations:

(*a*) The cultivation of a crop required man's constant, or almost constant, attention during the growing period: hence he had to be available in order to look after it; and not only for that reason, for if he wandered away and left it for a while someone else was likely to come along and take it.

(*b*) If man grew enough of his crop and carefully stored it, he had sufficient food until the next harvest and this made it possible for him to remain in one place.

Furthermore, we should recognise that man is by nature a gregarious creature, preferring to live in close contact with his fellow men rather than alone. Occasionally he has to live alone, or at least as a family unit, mainly because of economic conditions, e.g. the Scottish crofter, the Welsh hill sheep farmer. But usually he congregates in communities of various sizes, from the hamlet to the vast sprawling urban area.

3. Settlement types. Settlements fall into two main groups, rural and urban. In some cases, however, it is not always easy to say whether a settlement falls into the former or the latter category. For example, Portree on the east coast of the Isle of Skye is really a village in size and population yet it has quite unmistakably all the attributes of a town, and the same may be said of Hawes in Wensleydale, North Yorkshire. On the other hand, there are some towns which are essentially rural in their character, e.g. Northallerton, Bakewell, Selby.

Settlements are of varying kinds but seven main types can be distinguished:

(a) The isolated dwelling.
(b) The hamlet.
(c) The village.
(d) The market town.
(e) The town or city.
(f) Ribbon settlement.
(g) The conurbation.

The first four of these categories are rural settlements, with which we are concerned in this chapter. (*See* VIII for the remainder.)

4. The isolated dwelling. The farmstead, the manor house or large country residence, and occasionally the inn are the most common types of isolated dwelling. Farmhouses are commonly found in isolation, or relative isolation, lying as they do in the midst of fields which constitute the farm. Isolated farmsteads are found all over the world: the crofts in the Highlands of Scotland, the livestock farms of the Pennines, the prairie wheat farms of Canada, and the *estancias* of the Argentinian pampas are examples. Such settlements may be frequently many kilometres apart, hence, to a very large extent, they must be self-sufficient. The country residence, such as a hall or mansion, is frequently an isolated building, although it may lie just detached from the neighbouring village. Occasionally an inn may lie in isolation, especially if it is situated on a much frequented route; there are a number of such inns on the trans-Pennine roads. Usually the isolated dwelling is the product of economic necessity rather than social preference.

5. The hamlet. The distinction between the hamlet and the village is not clear-cut. Typically, however, the hamlet is smaller, usually

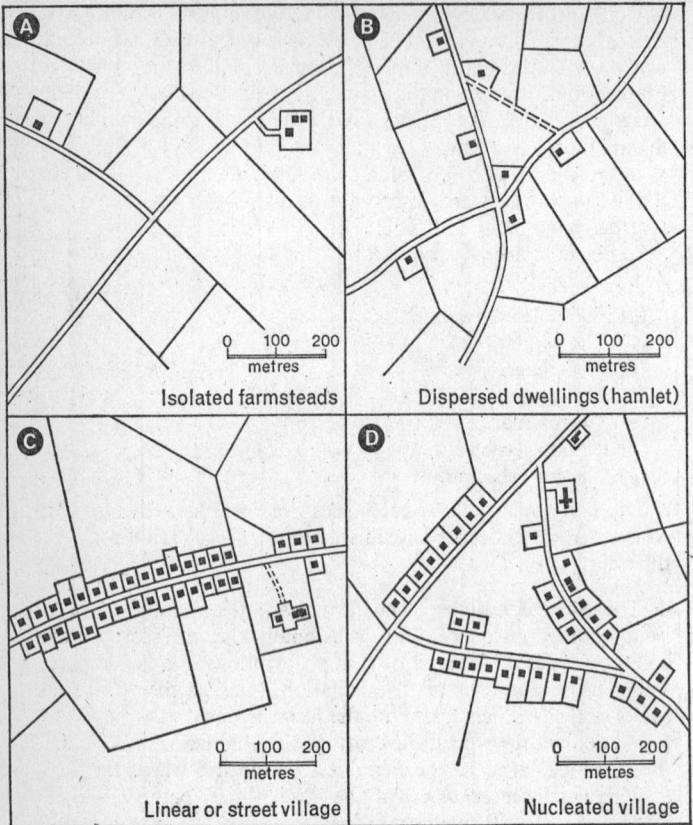

FIG. 14 *Settlement patterns.*

This shows the different kinds of settlements: (A) Isolated farmsteads:
(B) A loose grouping of dispersed dwellings forming a hamlet: (C)
Houses strung out along a road in linear fashion producing a "street"
village: (D) Dwellings built around a green to form a fairly compact
settlement.

much smaller, and the buildings composing it more scattered.
The presence or absence of a church, an inn, post office or school
is no criterion of hamlet status; although a hamlet usually does
not possess any of these social features. On the other hand, one

tiny, beautiful hamlet, Hubberholme in upper Wharfedale in North Yorkshire, virtually comprises three buildings: a church, an inn and a farm. In Britain, hamlets, like isolated farms, are usually associated with livestock rearing in hilly areas; they are more typical of Wales and Scotland than of England.

6. The village. Villages are of variable size; they may have anything from a few dozen to over a thousand people. They may be fairly compact or spread out over a considerable area. Typically, the village has certain social features, e.g. church, inn, school, store with post office, village hall, etc., though each of these is by no means always to be found; there are cases of quite large villages having neither church nor inn. But it is the presence of these social features which normally distinguishes the village from the hamlet. The small village of Stoke Abbott in Dorset, which has not much more than one hundred inhabitants, has a church, an inn, a village shop, a post office and a village meeting room.

The farming village is the oldest, as well as the most characteristic, village type; but there are, also, fishing villages and mining villages, the last being almost invariably a product of the Industrial Revolution. The coalfields of Durham, South Yorkshire and South Wales have numerous villages which developed through coal-mining activities.

7. The market town. Villages located in areas of good farming land and advantageously placed for communications usually became nodalised; in other words, they become foci of routes. Because of their function as marketing centres, they grew in size, often developing into market towns, e.g. Market Weighton in Humberside, Marlborough in Wiltshire, Aylesbury in Buckinghamshire. While some small towns may have grown up for alternative reasons, e.g. fishing, quarrying, mining, manufacture, communications, most British townships have their origins in local agricultural marketing centres. This, in fact, is true of towns throughout the world.

FACTORS AFFECTING SETTLEMENT

8. Settlement sites. Settlement is not a haphazard business. One important fact to note is that settlements, whether large or small, have sites which are due to the working of fundamental geographical laws. Seldom is a settlement site an accident, although there are some cases of this, e.g. certain settlements which grew up because of ecclesiastical associations. Sometimes these days "new towns" are "planted", but even in these cases the planners do not locate the place at random: in other words, just any site will not do. In earlier days, as for example in medieval times, communications were very difficult and communities were very largely self-supporting and they chose localities which enabled them to be as self-sufficient as possible. Our ancestors, more alive than we are to the basic requirements of life, had a greater appreciation of the countryside and usually used its resources to the best advantage.

9. Water supplies. First and foremost man had to have access to a supply of pure drinking water. Water is a primary human need and it may be taken for granted that man would be most likely to settle near a sure supply. This is most obviously illustrated in desert or semi-arid lands where people congregate around springs or wells. Settlements based upon such water resources are called wet-point settlements. There are many examples of this type of settlement in England, for the "spring-line settlements" often found at the base of a scarp slope, as around the western edge of the Yorkshire Wolds and around the Lincoln Heights, fall into this category.

10. Farming land. After water, food is man's next basic need; hence sites offering fertile land for cultivation or good pasture for animals were looked for, and a locality which provided a varied assortment of farming land was much prized, e.g. water meadows for cattle and geese, arable land for the plough, and rough grazing for sheep, together with a coppice for timber and fuel. It should be noted that many old parishes and farms have elongated shapes and run transversely across valleys so that they extend from the valley bottom, the river itself usually forming the boundary, across the terraces up the hillsides to the summit (*see* Figure 15).

11. Dry land. In areas where the land was liable to be marshy or damp, or suffered seasonal inundation through river floods, up standing sites were chosen. Some of the oldest village settlements in the world occur in the Tigris–Euphrates lowland of Iraq but, because of the great floods which these rivers formerly brought to the plain, man built his settlements upon slightly raised platforms. A similar example is found in the Low Countries where the *terpen*, or mounds raised above the flood lands, formed the

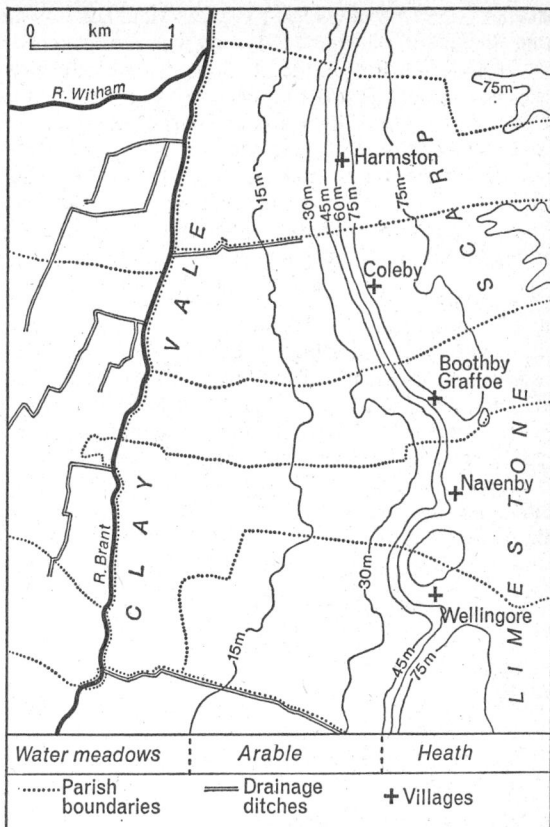

Map legend:

Water meadows	Arable	Heath

.... Parish boundaries   ═══ Drainage ditches   + Villages

FIG. 15 *Elongated, valley to hill-top parishes.*

In many parts of Britain parishes are long and narrow in their shape. This, usually, was the result of man's desire to procure varied types of land: meadow land for his cattle and geese; arable land for crop cultivation; and rough grazing and coppice for his sheep and pigs and fuel.

The scarplands, in particular, show this elongated parish pattern.

sites of the earliest settlements, e.g. Antwerp (Flemish: Antwerpen). In the English Fenland, Ely and other early settlements occupied elevated sites; these are called dry-point settlements.

12. Shelter. Other things being equal, man would choose a sun-catching site and one sheltered from cold winds or clammy river mists. Very often, it is true, the other factors, which were more necessitous, outweighed this consideration. But in many upland areas, e.g. in the Yorkshire Dales and in the Alps, settlements tend to be located on the sunnier, southward-facing hill slopes or valley sides. Exposed and wind-whipped sites will be avoided if possible and there are innumerable examples of English hamlets and villages which are tucked snugly away in hollows and little valleys to avoid the cold east and north winds.

13. Defence possibilities. In former times security from attack by hostile neighbours, marauders and pirates was often a necessary condition with respect to siting: hence man sought out defensible sites such as hill-tops, rocky outcrops, peninsulas, river loops, islands, etc., all of which lent themselves to easy defence. Defensive sites of this kind are sometimes called fortress points. In Italy, for instance, many of the villages occupy hill-top sites, while in southern Italy man congregated in villages rather than in individual farmsteads for reasons of security, for piratical attacks were common. Moreover, the villages were sited inland from the coast in order to avoid sudden attack from enemy forces by sea.

VILLAGE FORMS

14. The form of the village. The form or plan of the village and the pattern which its dwellings take is very variable, but one or two particular forms recur to give us fairly standard types: these may be categorised as follows:

  (a) Square or round villages.
  (b) Linear villages.
  (c) Crossroad villages.

These three basic forms are widely spread and have an almost world-wide distribution but they are most common in well-populated agricultural areas. There are, of course, many variants and some village forms are combinations of the above types.

15. Site influences. The form of the village has often been closely adapted to the natural features and characteristics of the site. Generally speaking the village is as compact as the terrain permits. The influence of the topography is clearly seen:

(*a*) In many hill-top villages where there is a concentric pattern of houses and streets.

(*b*) In river-side villages or canal villages which are strung out along the bank.

(*c*) In narrow valleys hemmed in by steep slopes where the village threads its way along the valley bottom.

(*d*) In creek settlements where the houses are clustered around the opening and run inland by means of steep and narrow streets.

16. The square or round village. "There are many variations," writes D. C. Money (*Introduction to Human Geography*, U.T.P.), "but in general most of the houses surround a green, pond, market place, or church compound. In later times they clustered around the entrance to a great house, as at Kimbolton, Huntingdon." There are many examples of villages which consist of a green surrounded by houses: good examples are common in Yorkshire, e.g. Bainbridge in Wensleydale and Reeth in Swaledale. In some cases the present form of the village reflects its earlier shape provided by its protective stockade: for example, in northern England and the border country one can see villages which are quite clearly derived from the old square stockade, e.g. Blanchland in Northumberland, while in eastern Europe the *rundling*, a "round" village, taking its shape from its original ring-fence, is a very common type (*see* Figure 16).

17. Linear villages. These elongated villages, which sometimes straggle over long distances—sometimes as much as two kilometres or more—usually grew up in the following ways:

(*a*) Along lines of communication:
    (*i*) Along a road.
    (*ii*) Along a river.
    (*iii*) Along an artificial waterway, such as a canal.
(*b*) Along constricted valleys or on riverside terraces.

The *strassendorf* or "street" village, which occurs in many parts of north central Europe, is commonly associated with the penetration and clearance of the forest lands in medieval times. Linear forms are also associated with marshlands and the polderlands of the Low Countries and the North German coast. Villages along riverside terraces or on high shelves, such as Feetham and Low Row in Swaledale, take elongated forms. Some villages in the Pennine upland are long winding settlements, almost one house deep, because the relief does not allow development in depth.

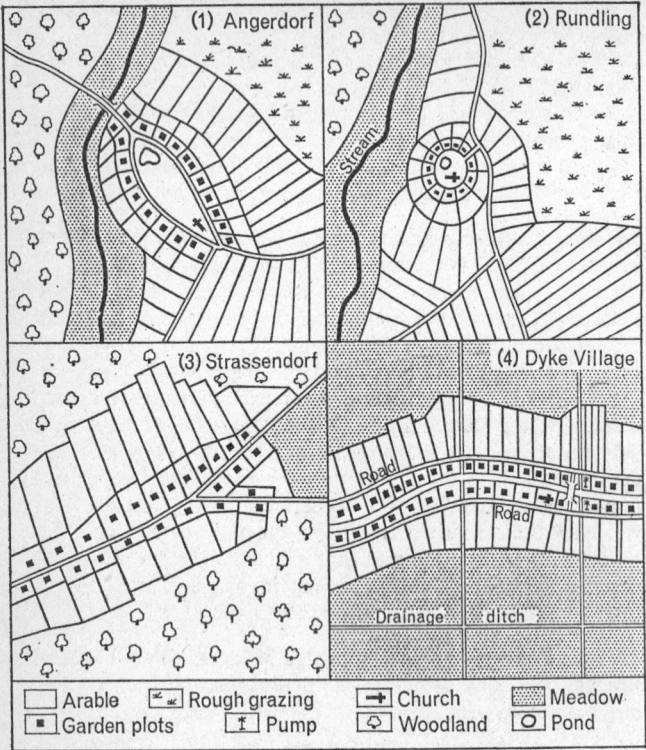

| Arable | Rough grazing | → Church | Meadow |
| Garden plots | Pump | Woodland | Pond |

FIG. 16   *Types of European villages.*

Villages show a variety of shapes and forms but certain types are of common occurrence in different parts of Europe. (1) The *angerdorf* is a type of street village built around an open, oval-shaped space or green: (2) The *rundling* or round village (*runddorf*) is common in what is now East Germany: it is centred on a village green: (3) The *strassendorf* or street village is associated with colonisation in, and clearance of, the forest: (4) The dyke village takes a linear form since it was built along the dykes in the empoldered lands.

18. Crossroad villages. There are numerous variants of this form, depending upon the angle made by the converging roads. One may suggest three simple forms:

(a) A Y or T junction; where one road joins another at an acute or right angle.

(b) A true crossroad, where one road cuts another more or less at right angles.

(c) A multiple junction where several roads join to form a "star".

PATTERNS OF RURAL SETTLEMENT

19. Basic patterns. If we study maps of rural areas we find that there are two main types of patterns of rural settlements; they are either dispersed, i.e. scattered about the countryside, or nucleated, i.e. concentrated together into fairly compact groups.

These different patterns pose a number of questions: Which pattern came first, the dispersed or the nucleated? Can a dispersed pattern change to a nucleated pattern, or vice versa? Which factors decide whether the pattern shall be dispersed or nucleated? These questions are discussed in **20–22** below.

20. The primary pattern. It is not easy to look back into the past and it is dangerous to generalise but such evidence as there is tends to suggest that the original pattern was nucleated. Man, as we have already said, is a gregarious creature who has a distinct proclivity for remaining in groups whether it be the family, clan, tribe or nation. In simple societies the cohesion of social groups is demonstrated by the communal dwelling, e.g. the long houses of the Dyaks of Borneo which contain as many as thirty or forty families; the long house literally constitutes the village. On these grounds, as Emrys Jones writes, "theoretically one might put forward a case that the small nucleated settlement is the primary form, and that all dispersal is subsequent" (*Human Geography*). One can appreciate how, in early times, protection and security were primary needs and that safety could be best ensured by living in close contact with others.

21. Changing patterns. Research, particularly by French scholars, has shown that dispersion and nucleation within a given area may change according to the conditions which operate at different times: in other words, the exigencies of one period may lead to an essentially nucleated pattern of settlement giving way to dispersion while, at another, conditions may result in dispersed settlements being replaced by nucleated villages. For example, during periods of social insecurity, such as the fifth, sixth, tenth

and early eleventh centuries A.D., the disturbed conditions resulted in the disappearance of numerous isolated, dispersed
farmsteads and the congregation of peasants into villages for
defence. Again, the ravages of the Black Death and the shortages
of labour consequent upon the Hundred Years War in the fourteenth century also resulted in the disappearance of many isolated settlements. On the other hand, whenever colonisation or
enclosure of wasteland took place the settlement which emerged
tended to be dispersed in character. Periods of peace, law and
order also tempted peasants to leave the village and to increase
the extent of dispersion. J. M. Houston (*A Social Geography of
Europe*), quotes how, in the fifteenth and sixteenth centuries in
France, the devaluation in money and the associated drop in
value of fixed village rents compelled the landowners to encourage
tenant settlers to take up isolated holdings on their estates.

22. Influencing factors. Which of the principal rural settlement
types—single dispersed farms, scattered hamlets, nucleated villages—predominates in any area depends upon a variety of
natural and human factors. They cannot be explained in terms of
any simple physical relationship; complex historical, social and
economic influences are often also involved. Amongst the various
influencing factors at play may be the following:

(*a*) *Water supplies.* Shortage of surface water supplies may
lead to the concentration of settlements beside a stream, at an
emergent spring, or around a well; conversely, abundant surface
supplies may cause dispersion.

(*b*) *Rugged terrain.* The difficulties of tillage in rough, upland
country frequently result in the dispersion of farmsteads which
concentrate on livestock rearing; concentrated settlement tends
to occur in the valley bottoms or in patches of lowland.

(*c*) *Peace and security.* Periods of peace, law and order were
favourable to the outward dispersion from an original nucleated
settlement. On the other hand, unsettled political and social conditions, giving rise to insecurity, tended to cause people to come
together for common defence.

(*d*) *Economic influences.* As was noted above, the devaluation
of money and the enclosure of wasteland (which happened in
England in the sixteenth century for sheep grazing and wool production) had the effect of increasing the extent of dispersion.

(*e*) *Social influences.* These, as for example, customs of inheritance whereby land was shared by the sons of a landowner

or where a parcel of land was given as a dowry, were likely to lead to the fragmentation of land and the dispersion of settlement.

(*f*) *Historical influences.* Peoples invading and settling territory or colonising land were apt to introduce the forms of settlement to which they were accustomed; for example, the Anglo-Saxons, who were village dwellers, created villages when they colonised the English lowland.

DWELLINGS

23. Shelter as a primary human need. Although man's first need is for food and water, the necessity for protection and shelter is also a requisite for existence and is universal. Man needs a measure of protection and security from the following hazards:

(*a*) Climatic extremes.
(*b*) Preying animals and insect pests.
(*c*) The hostile actions of his fellow men.

Because of these reasons man has devised, in varying degrees of complexity, some form of shelter and protection. The dwelling he has devised is a distinctive cultural feature of the landscape and whether singly or in groups is a topic for geographical study.

24. Types of shelter. Man has always sought some kind of shelter and no doubt the earliest was the cave which provided a convenient, ready-made home. For example, we know that the caves in Creswell Crags, in Derbyshire, were occupied by Palaeolithic Man 10,000, perhaps 15,000, years ago. Caves, however, were not always available and so man had to find other forms of shelter. Primitive man, partly because of his limited technology, often had to be satisfied with a mere windbreak but "through degrees of elaboration men have devised simple conical huts, tents, log dwellings, mud and brick and stone structures; some to house families, others to serve society in a wider sense. Differences in the form and function of these structures vary enormously" (E. Jones, *Human Geography*, Chatto & Windus).

25. Varying demand. The needs of shelter, and therefore the degree of its elaboration, vary from region to region, largely as a result of the climatic conditions. In warm regions the house is less necessary than in cold lands; it can be a more flimsy structure and its main function is to provide shelter from the rain. Since it is warm, air circulation is necessary, and there are often

no windows. If it is very hot, houses will often have the minimum
of openings to exclude the heat. In hot, wet tropical lands dwellings will normally be raised off the ground and mounted on
stilts to give protection against damp and insect pests. On the
other hand, in cold, windy climates houses must be strong, warm
and snug to give adequate protection from frost, snow and bitter
winds.

26. An example of contrasting conditions. The house in the
Mediterranean lands plays a less important role in social life
than it does elsewhere in Europe. Much of the daily life of the
individual in the Mediterranean countries is spent in the open
air, in the patios, in the streets, in the squares, in the market-
place; he uses his home largely as a resting and sleeping place
and as a store. Social life tends to be focused on the streets, in
the sidewalk cafes, in the open squares, or around the campfire
and not, as in Britain, in the home. Generally speaking, houses
are smaller, with fewer rooms than ours, not so much because
the people are poorer and have a lower standard of living, but
simply because houses are a less necessary social need.

27. Relationship between house and habitat. The plans, shapes,
construction and building materials of dwellings are often ex-
pressive of the relationships that exist between man and his environment. Although houses differ in their architecture, planning
and building materials according to the taste and wealth of their
individual owners, dwellings in general throughout the world
show definite adaptations to regional climatic conditions and
responses to local natural conditions and resources. This relationship was more apparent in the past than at present, for the
spread of cultural features, the relative ease of transportation of
materials, the use of standardised products and the use of synthetic building materials are tending to change regional house
types.

28. English regional types. While housing estates are now being
built almost everywhere in England and the above mentioned
factors are tending to undermine specialised regional building,
a tour of the country shows how, in the past, the dominance of
one building material or another in a particular area gave rise to
distinctive regional building types:

(*a*) The timber frame, half-timbered, black and white houses
of the Welsh border country.

(*b*) The warm, mellow buff-coloured limestone cottages with their stone roofs of the Cotswold country.

(*c*) The timber frame and red brick houses, capped by red pantiles or thatch, of much of the Midlands.

(*d*) The flintwork which became a distinctive feature of East Anglia where good building stone was practically non-existent.

(*e*) The solid, squat and snug gritstone cottages with their stone slab roofs of the Yorkshire Pennines.

29. Geographical influences. Let us conclude this brief account of house types by listing a few fairly obvious examples of the influence which the environment has had upon building.

(*a*) *Hilly, broken terrain.* Steep slopes and a scarcity of flat land often lead to tall, many-storeyed houses with small ground plans:

(*i*) The houses of the hilly coastal areas of the Gulfs of Genoa and Salerno.

(*ii*) The mud-brick skyscrapers found in the Yemen and the Hadramaut.

(*iii*) The three- or four-storeyed dwellings built on many of the steep scarp edges in many parts of the Pennines.

(*b*) *Heavy rains and snows.* In some temperate and tropical regions heavy precipitation has resulted in high-pent roofs and overhanging eaves:

(*i*) The steep roofs of Norwegian houses to shed the copious rains and snows.

(*ii*) The steep, heavily thatched roofs of native dwellings in India with its monsoon and tropical rainstorms.

(*c*) *High temperatures and sunshine.* Roofs are flat or low pent while wall apertures are few and small. Balconies, colonnades and cloisters are features giving welcome shade during the heat of the day:

(*i*) The patio, a shaded quadrangle, is a feature of many Spanish houses.

(*ii*) The dwellings in the Middle East are box-like in shape with flat roofs and a minimum of wall openings.

(*iii*) Verandahs are common features of houses in hot, sunny lands, e.g. India, the southern United States.

(*d*) *Strong winds.* Regions experiencing strong, persistent winds or sudden air movements tend to show dwellings with stout, heavy roofs, while doors are placed on the sheltered side:

(*i*) In many parts of Austria and Switzerland the heavy

chalet roofs are "anchored" by great boulders to withstand the fierce blow of the föhn wind.

(*ii*) In parts of Ireland and Wales thatch is tied and pegged and roof tiles are cemented to guard against high winds.

(*iii*) In Brittany and Flanders houses are built with their fronts facing east or south-east, never to the north-west, again to avoid exposure to strong winds.

(*e*) *Building materials.* These influence the structural forms of buildings through their inherent architectural possibilities and limitations.

(*i*) In Finland and eastern Canada houses are built preponderantly of wood since timber is abundant.

(*ii*) In the flat, alluvial lowlands of the Netherlands where building stone is absent, houses are of brick and wood.

(*iii*) In the thickly forested lands of south-east Asia where vegetable material is present in abundance, dwellings are of wood, bamboo, rattan and thatch.

PROGRESS TEST 7

1. Which developments enabled man to live a permanently settled life? **(1, 2)**

2. Describe the chief differences between the hamlet and the village. Quote examples you know of each. **(5, 6)**

3. Which are the chief factors affecting the siting of settlements? **(9–13)**

4. Explain the meaning of "wet-point" and "dry-point" settlements. Quote examples of each. **(9, 11)**

5. Explain how geographical conditions may affect the shape of village settlements. **(15, 17)**

6. Which factors may help to explain the predominance of a particular settlement type? **(22)**

7. Why is shelter termed a primary human need? **(23)**

8. Why are warm, snug houses less necessary in Mediterranean lands than in the British Isles? **(26)**

9. In what ways are house types adapted to environmental conditions? **(28)**

10. Illustrate how the following have influenced building: (*a*) great heat and sunshine, (*b*) strong winds and cold, and (*c*) heavy rains and snows. **(29)**

CHAPTER VIII

Urban Geography

TOWNS AND CITIES

1. Definition of a town. Towns may be distinguished from villages by the fact that the number of non-agricultural workers exceeds the agricultural workers. Hence the number of persons employed in non-agricultural pursuits is one simple way of deciding whether a particular settlement should be classed as a village or as a town.

Villages of any size will have a village store, a post office, a garage and a bank (even if it is open only on certain days), but the shopping, professional and transport "services" will be limited. Towns, on the other hand, will possess a wide variety of "services": they will have many shops, specialising in the sale of particular goods, as distinct from the general village store; they will have offices, banks, schools, libraries and places of amusement; and they will have varied and fairly frequent transport services both within the town and to outlying villages and nearby towns. In other words, a number of specialised activities and services are to be found in towns which normally are absent in villages. Thus, a town may also be distinguished from a village by its function, i.e. what the town does or why it exists.

Size is usually a distinguishing feature, although size is not always a sure sign of town status. Size in relation to town status is quite an arbitrary matter: many places in Australia which the Australians call towns would, from the point of view of their size, be termed mere villages by an Englishman; on the other hand, in China there are many villages which in terms of their population numbers might, feasibly, be called towns.

2. Town and city. The terms "town" and "city" are loosely used and in general are used interchangeably. If any distinction is made by the man in the street, the term city is usually meant to imply a very large town. The word city has, in fact, various connotations:

(*a*) To some people, particularly Londoners, city is popularly used to mean the old, central part or nucleus of a town, in contrast to the more recent outer and suburban growth. On the continent, and particularly in France, city has precisely this meaning. When Parisians speak of the *cité* they mean the island in the River Seine which was the site of the original settlement of Paris.

(*b*) The term city in the ecclesiastical sense refers to a town which is, or has been, the see of a bishop; in other words, city means a town, often though not necessarily an old-established town, with a cathedral and a bishop. Such towns need not be large; some in fact, such as Ely, Wells and Ripon, are quite small.

(*c*) In the United Kingdom, however, the term city does have a specific meaning: it is a title created by royal authority. Since the latter part of the nineteenth century the official style of city has been conferred upon certain important towns through the granting of a charter of incorporation by royal authority. Birmingham was the first town to be distinguished in this way; it became a city in 1889. The possession of such a title, however, confers no specific privileges other than ceremonial ones.

3. **Towns as dynamic organisms.** Towns are living units and, like people, are born, grow, decline and may die; they may also have a distinctive character or personality. Towns never remain static: they are constantly changing in their shape, size, plan, architecture and function. This becomes evident if you re-visit a town after several years. The amount of change and the kind of change vary, of course, between one town and another but all in time undergo change. Some examples of towns which have changed over the years are as follows:

(*a*) *Hastings* was once an important port, the leading town in the confederation of the Cinque Ports; today it is mainly a seaside resort.

(*b*) *Chester-le-Street*, County Durham, was originally a Roman fort, became the see of a bishop, and is now a colliery town.

(*c*) *Peterborough*, in Cambridgeshire, began as a monastic and ecclesiastical centre, grew to be an agricultural marketing town and then, with the coming of the railway, grew into a busy railway centre and subsequently into an industrial town.

(*d*) *Ilchester*, in Somerset, originated as a Roman military station, became a thriving merchant centre in medieval times and the administrative centre of the county but has now dwindled to a village.

(*e*) *Ilkley*, occupying the site of a Roman fort, developed as a spa in the mid-nineteenth century but is now largely a residential suburb of Bradford and Leeds.

(*f*) *Old Sarum*, a Roman, Saxon, Norman and medieval town which lingered on until the seventeenth century, no longer exists. Salisbury, 20 km away, came to replace it.

4. The personality of towns. Although some towns, such as many seaside resorts or some of the Yorkshire wool textile towns, are similar, no two towns are exactly alike. This is because of the following reasons:

(*a*) The peculiarities of their physical sites.
(*b*) Their different histories.
(*c*) Their roles and functions.
(*d*) Their plans and architectural features.

These help to give a town a character or personality. Some towns, such as York, Brighton, Bath, Cheltenham and Norwich, possess personality to a marked degree.

A geographer once wrote: "There are newborn baby towns; doddering old-man towns; towns in a hectic whirl of youth; poor, guttersnipe towns; fat, millionaire towns; quiet, studious towns; loud, blatant towns; towns with all the luck, and towns with no luck at all; dying towns; dead towns. Towns are very human" (H. Alnwick, *A Geography of Europe*). This somewhat light-hearted and amusing description contains much truth and the writer had a deeper and more subtle understanding of the growth and character of towns than his words might, at first glance, suggest.

THE GROWTH OF TOWNS

5. The urban revolution. The idea of urban life, i.e. the concentration of people in towns, is usually said to have taken place during the fifth millennium B.C. in the great river lowlands of the Near and Middle East (*see* Fig. 17). However, recent excavations at Jericho, in the Jordan valley, indicate that there was an urban settlement here in the seventh millennium B.C. and so we may have to reconsider the date when towns first appeared. The custom of living in towns, known as the urban revolution, had at all events become customary by the beginning of the historical period.

The idea of urban life started in the East but gradually spread throughout the Mediterranean lands. The Minoans, Phoenicians

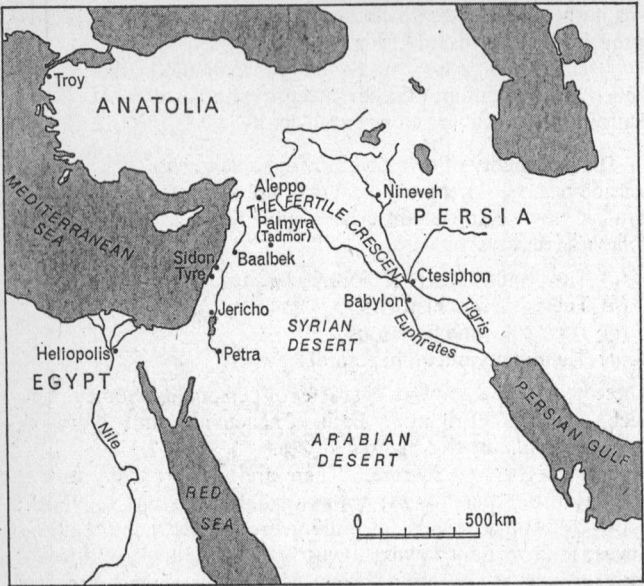

FIG. 17 *Ancient towns of the Near and Middle East.*

In this arid and semi-arid part of the world settlement clung closely to water supplies and to the rivers and streams which provided water for irrigation. A series of trading towns developed along the "Fertile Crescent" which linked the Persian Gulf with the eastern Mediterranean.

and Greeks all built towns. Later, the Carthaginians and Romans continued the practice of colonisation and urban settlement.

6. The importance of urban development. Urban development represented a cultural advance beyond the stage of the first nucleated settlements which were villages. With urban growth were associated various new inventions and practices:

(*a*) *Architecture and the building of structures,* other than houses, which served society in a wider sense, e.g. temples, tombs.

(*b*) *Metallurgy* which led to the replacement of stone weapons, implements and tools by first copper and then bronze artifacts, which were far superior.

(c) *Organised trade* which led to the specialisation of labour, the development of transport, and a more sophisticated life.

(d) *The art of writing*, largely an outcome of trade and the need for keeping business records, which led ultimately to literary expression.

These together, according to the late V. G. Childe, ushered in the second great cultural revolution in human history. Towns were the result of developing human needs for security, trade or welfare.

7. Trade as the raison d'être of urban development. The growth of business, trade and exchange was probably *the* fundamental factor in urban development, or *raison d'être*. Some towns, it is true, emerged as defensive, administrative and ecclesiastical centres but in all probability the great majority grew as trade centres. Subsequently towns developed to perform specialised functions and many developed, also, as centres of specialised craft industries. In time, towns became multifunctional.

8. Civic life. The growth and activities of towns led to the accumulation of wealth, and wealth provided leisure: leisure, in turn, encouraged developments in art, literature and music and in the realm of thought. The complex of ideas which sprang from urban life led to the growth of what is called civilisation, i.e. civilised life with all the refinements that are implied. Civilised living contrasted strongly with the primitive and barbarous life of unsettled peoples.

9. The Romans as town builders. W. G. East wrote: "The town was indeed the hallmark of Romano-Greek culture" (*A Geography of Europe*). The Romans continued the Mediterranean tradition of town building and the first real towns north of the Alps owed their origin to them. Such towns developed from the *castra* and *colonia*—their military camps and their colonial settlements. When the Romans conquered England they developed London as a civic capital from which roads were built in all directions, like the spokes from the hub of a wheel, and along these highways garrison towns were established at, for example, Dover, Colchester, Lincoln, York, Chester and Gloucester.

The Romans showed remarkable prescience in the siting of their towns: they chose strongpoints which guarded strategic routeways and selected nodal points which were foci of land and water routes commanding trade. So long as the *Pax Romana*

prevailed these towns flourished but the collapse of the Empire
led to the decline, and sometimes the extinction of these urban
centres.

> NOTE: After the Barbarians had overrun the Roman Empire
> and had settled, they built scattered villages and started to
> cultivate the land. As agriculturalists they were essentially
> village, not urban, dwellers. The local, rural communities set
> up by the Barbarians were mainly self-sufficing units and they
> had little need of exchange: hence trade languished and they
> had no use for the towns which had been built by the Romans.
> In fact, the Barbarians were more ready to destroy the towns
> than inhabit them.

10. Medieval towns. Many of the towns established by the
Romans, though perhaps sacked and reduced in population, did
survive and slowly recovered. Again, as time went on, new cities
arose; chiefly, they owed their origin to three causes:

(*a*) Many resulted from the concentration of population for
purposes of security around castles, e.g. Edinburgh, Warwick,
Carcassonne, Ghent.

(*b*) Some grew up around monasteries and episcopal seats, e.g.
Canterbury, Salisbury, St. Andrews, Tours, Turku (in Finland).

(*c*) Others grew up in connection with trade and commerce.

(*i*) Ports such as Bristol, Kings Lynn, Bruges, Bordeaux,
Venice.

(*ii*) Route foci, bridging points, etc., such as Northallerton,
Oxford, Basle, Coblenz.

11. The rise and decline of towns. As we mentioned in 3 of this
chapter, towns are living organisms and dynamic centres. Meta-
phorically speaking, they thrive and grow if they are fed and they
wither and die if they are starved. Urban expansion and prosperity alternate with impoverishment and decline. Several factors
are at play in causing such fluctuations:

(*a*) Natural disasters, such as earthquakes, volcanic eruptions,
storms, floods, coastal erosion, pestilence, fire, etc., may destroy
cities, e.g. Lisbon, Ravenser Odd (off Spurn Point), Pompeii.
Often, however, if their situations are of major significance, new
towns will appear on or near the old site. Think of the thirteen
cities which have arisen on or near the site of present-day Delhi
which clearly indicate the great strategic importance of its situation.

(*b*) Historical and political factors may explain some changes. A town may be destroyed, like Carthage and Wisby, as a result of war, or decline, like Trieste, which was a result of the boundary change in 1918 which cut off the port from its natural hinterland. Madrid and Leningrad were artificially created as capital cities of Spain and Russia respectively. An old, small, provincial town may suddenly, as the result of political factors, become important and expand, like Bonn, the present capital of West Germany.

(*c*) The site value may alter with the passage of time. The potentialities of Lyons lay dormant until the Romans brought the Rhône–Saone corridor under their control and opened up the valley as a routeway. Lisbon, situated on the Atlantic coast, began to flourish with the development of deep-sea oceanic trading. The medieval town of Kiel, at the base of the peninsula of Jutland, received a new lease of life with the cutting of the Kiel Canal.

(*d*) The decline of trade routes, hitherto of great importance, may lead to the towns along such routes sharing in the decline, e.g. Petra and Palmyra in the Near East. The Italian merchant cities of Venice and Genoa declined when the ocean routes to the East were opened up, as also did some of the trans-Alpine trading centres such as Augsburg. The development of new routes, on the other hand, stimulates growth, e.g. the Suez Canal and Port Said.

(*e*) Ports may be seriously affected by silting, or the increase in the size of vessels, or the changed economy of the hinterland. Aquileia, a flourishing port in Roman times, now lies inland and is a place of no consequence. Bruges, once the most important port in northern Europe, is now a quiet backwater. Many estuarine ports, such as Hamburg, Bremen and Bordeaux, have been compelled to create outports in order to maintain their importance.

(*f*) A natural resource may be discovered, exploited, and give rise to new towns, e.g. Corby (based on iron ore) in the English Midlands, Genk (coal) in the Belgian Kempenland, Sidi Kacem (petroleum) in Morocco. Conversely, a resource may become exhausted or of reduced value and, accordingly, towns dependent upon its exploitation may decline as, for example, the mining "ghost towns" of the western United States, many of the west Durham coal-mining centres and the North Wales slate-quarrying towns.

THE SHAPE AND STRUCTURE OF TOWNS

12. Town form. Every town is an individual place with its own particular character, as we have already stressed. The individuality arises from the town's site and situation, its historical development, its particular activities, and its architecture and planning. All this expresses itself in the form of the town, which embraces the following:

(*a*) Its shape, whether regular or irregular, round, square, elongated, etc.

(*b*) Its structure, the internal organisation or zoning which prevails.

(*c*) Its skyline, or silhouette.

Although every town is unique, the majority of towns possess certain features and elements which are shared in common.

13. Shapes of towns. Broadly, towns are either regular or irregular in their shape. If they are "planned" towns they tend to have a regular shape; if unplanned they are usually irregular.

The Romans were an essentially practical people and were amongst the first to plan towns. Often the towns were based upon, or grew out of, the *castra* or military fort which typically was square, walled and had a grid plan of roads. Many towns of Roman origin exhibit to this day the rigid and formalised plan or pattern of the original settlement; e.g. Arles and Nîmes clearly show this planning and it can be discerned in Chichester.

Although most medieval towns show little, if any, planning, a considerable number were in fact planted as planned settlements; for example, in England Flint and Hull still show some evidence of their original planning. An interesting early example is St. Albans where in the middle of the tenth century Abbot Wulsin laid out a rudimentary plan; this simple plan exists today.

Some fine examples of *bastide* towns, or towns of military origin, occur in France. The rectangular shape, planned layout and fortifications of Aigues Mortes, in the Rhône Delta, are remarkably well preserved and there has been very little extension of settlement outside its walls.

Many towns, and especially the industrial towns of modern times, grew quickly and in haphazard fashion with the result that they are often highly irregular in shape, e.g. Huddersfield, Stockton-on-Tees, Swansea.

Seaside resorts are often semi-circular in shape because of the importance of the sea frontage.

14. Urban structure. The morphology or structure of towns is an important aspect of urban geography. Most towns have distinctive areas or neighbourhoods—shopping centres, industrial areas, zones of old and decaying property, suburbs. In an attempt to explain the dynamism and the functional zones of towns a number of urban geographers have advanced theories of urban growth and functional development.

(*a*) *The concentric theory.* This theory, that towns tended to grow outwards from an original core or nucleus and in time developed an annular structure (one of concentric rings), was suggested by R. E. Parks and E. W. Burgess, over fifty years ago. Burgess had made a study of the lakeside city of Chicago and found a number of distinctive regions. He suggested that any large city tended to expand outwards and that a series of zones developed, each growing by gradual colonisation from the inner zone next to it. As one travelled inland from the waterfront, there were five main zones: (*i*) a central or core area, known in Chicago as "the Loop", which was the Central Business District (C.B.D.); (*ii*) a surrounding semi-circular zone of decaying property and slums with commercial offices and minor manufacturing industries; (*iii*) beyond, a zone of factories and working-class houses; (*iv*) beyond still, a belt of better-class houses in a suburban residential area; and (*v*) an outer high-class residential zone where commuters lived in desirable properties in a semi-rural setting. According to Burgess this kind of concentric arrangement was applicable to most "Western" cities whether they were in North America, Europe or Australia.

R. E. Dickinson, a geographical specialist in settlement studies, applied Burgess' theory to British towns; he discovered a similar structural pattern and suggested a basic division into four urban zones: (*i*) the central or nuclear area, which was the old part of the town and which came to be the business area with its stores, offices, markets, warehouses, hotels and civic buildings, etc.; an area mostly non-residential; (*ii*) the middle zone surrounding the core which was a decaying zone of rundown property and often immigrant population; it was mainly a zone of late nineteenth- and early twentieth-century building with terraced dwellings, some shops, churches and chapels, warehouses and small industrial premises; (*iii*) the outer zone built up after the First World

War with the coming of improved communications (tramcar, buses and motor cars), a zone essentially of housing estates but with newer industrial plants growing along the main highways and by-pass roads; and (*iv*) the urban fringe, a zone of fairly open country punctuated by high-class residential areas, mainly of the post Second World War period, the new houses frequently congregating around older villages.

(*b*) *The sector theory*. H. Hoyt, who made a study of the growth and structure of residential neighbourhoods in American cities, postulated the idea of town development in terms of sectors. Like Park and Burgess, he recognised that towns had a core area or C.B.D. but visualised the residential areas as expanding outwards from the city centre to its periphery in the form or wedges or sectors. By studying the varying rents which were charged in different areas in the city and noting that they changed axially rather than concentrically, he was able to support his thesis of a sector pattern. Hoyt also recognised that some sectors might be given over to light manufacturing industry. One can easily conceive of a town located in a basin where streams flowing towards the centre of the basin would carve out a series of valleys separated by intervening ridges: the ridges would form attractive residential belts, the valleys, because they offered flatter land and facilitated communications, would be likely to attract warehouses and factories. Many towns in the foothill zone of south-west Yorkshire display this sector pattern quite clearly.

The city of Bradford, more especially before 1939, provided a good example of a town which exhibited both concentric zoning and sectors (*see* Fig. 18). The town, a small village around 1800, grew rapidly during the nineteenth century and prospered through its wool textile trade. The original settlement grew up in a basin where three becks joined and its name means "broad fording place". The old town grew up in the bottom of the basin and gradually expanded outwards and upwards towards the rim of the basin absorbing in its growth the townships of Bowling, Horton and Manningham. Before the Second World War, and before the re-planning of the city centre, the C.B.D. showed three fairly distinct and clear-cut zones: an area of wool warehouses and offices: an area devoted to banks, shops and markets; and an area largely devoted to entertainments and educational establishments. Around this central area was a zone of Victorian and Edwardian terrace houses—some of which originally were very splendid—but by the 1930s this was a decaying zone with

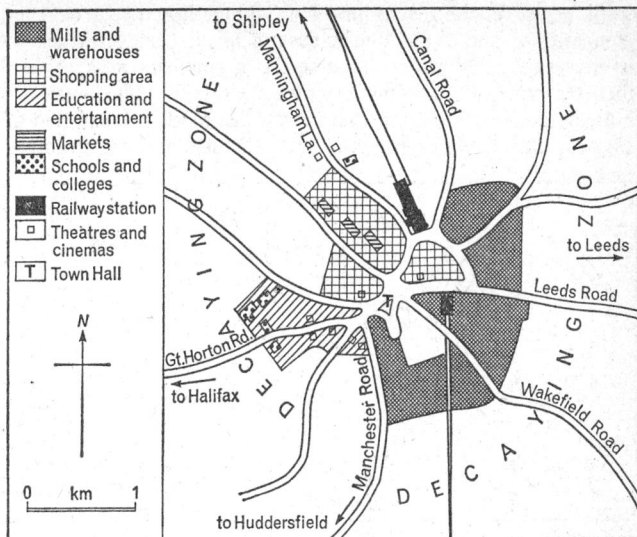

FIG. 18   *The structure of pre-war Bradford.*

developing slums. It is in this zone (now being cleared) that Bradford's large immigrant population of Indians, Pakistanis and West Indians has settled. In the zone beyond this extensive corporation and private housing estates were built. In post-war years newer suburbs have grown up largely around the outlying villages of Allerton, Cottingley, Apperly Bridge, Woodhall and Buttershaw which were largely located on the rim of the basin. But grafted onto this concentric zoning were industrial sectors which followed the lines of the Thornton, Horton and Bowling becks which flowed into Bradford basin. Along these valleys were mills, engineering works, dye-houses, etc.

(c) *The multiple-nuclei theory.* This theory, propounded by the Americans C. D. Harris and E. L. Ullman, recognised that towns might show both elements of concentric zoning and sectors but claimed that large towns normally contained several subsidiary centres whose growth complicated the expansion of a city from a central core. Harris and Ullman's theory can be applied to many large cities which, in their growth, have engulfed and absorbed nearby small towns, even though each of the latter may continue to function as minor foci within the large urban ag-glomeration. The most obvious case of this is London which as it grew swallowed up a whole series of smaller towns though these continue still to act a minor nodes within the metropolis.

It should be understood that these three models, illustrated in Fig. 19, are merely attempts to try to understand the structure

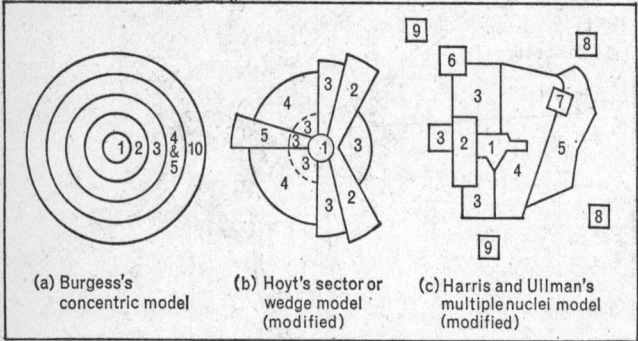

FIG. 19   *Models of urban structures.*

1. Central business area; 2. wholesaling and light industry (including slums); 3. low-class housing; 4. middle-class housing; 5. high-class housing; 6. heavy industry; 7. subsidiary business district; 8. outer suburban housing; 9. outer suburban industry; 10. wealthy commuters' zone.

and development of towns. Probably none of them fits precisely any actual city for each tends to have a morphology which is unique to itself. None the less, these urban concepts help us to understand the structure of towns and to trace and analyse land use regions within the city.

CLASSIFICATION OF TOWNS

15. Basis. Towns may be classified in a variety of ways: age; population; origin; function.

Each of these has its importance, but the geographer is especi-ally interested in the influence exerted by physical conditions on town development, and with the functions of urban centres.

16. Classification by origin. Both site and situation are a product of the physical environment.

(*a*) *Rivers.* A very high proportion of towns are located on rivers because in the early days they provided a water supply and gave the easiest means of communication. Particular points on a river frequently had special significance:

(*i*) A convenient fording point, e.g. Westminster, Wallingford, Oxford.

(*ii*) The lowest bridging point, e.g. Newcastle, Budapest.

(*iii*) The confluence of two rivers, e.g. Reading, Coblenz, Khartoum.

(*iv*) A bend where the river changes direction, e.g. Orleans, Timbuktu.

(*v*) Where a river meanders, forming a defensive loop, e.g. Yarm on Tees, Shrewsbury.

(*vi*) Where a river enters or leaves a lake, e.g. Geneva, Vänersborg (Sweden).

(*b*) *Gaps.* Where mountain ranges or highlands provide obstacles to movement, throughways or gaps channel the routes, and towns tend to grow up at the entrances and exits. These towns are called gap towns, e.g. Leeds, Guildford, Innsbruck, Kabul. Such towns may have great strategic and military importance.

(*c*) *Defensive sites* are provided by various features:

(*i*) An upstanding rock outcrop or isolated hill, e.g. Edinburgh, Athens, Salzburg.

(*ii*) A peninsula, e.g. Instanbul, Monaco.

(*iii*) An island or islands in a river or in the sea, e.g. Paris, Montreal, Venice.

(*d*) *Highland and plain junction:* here towns act as markets and exchange points for the produce of two differing regions, e.g. Perth, Besançon, Denver.

(*e*) *Centres of fertile lowlands and basins,* e.g. Norwich, Taunton, Rennes, Prague.

(*f*) *Route foci,* where natural land routes converge, e.g. Carlisle, Vienna, Chicago, St. Louis. This feature is often closely related to (*e*).

(*g*) Presence of a *natural resource,* especially a mineral resource, e.g.:

(*i*) Metals, e.g. Kalgoorlie (gold), Kiruna (iron), Chuquicamata (copper).

(*ii*) Fuels, e.g. Wigan (coal), Ploesti in Romania (oil).

(*iii*) Salt, e.g. Salzburg, Northwich, Château Salines.

(*iv*) Medicated water and thermal springs, e.g. Spa, Bath, Vichy, Saratoga Springs.

17. Classification by function. Many geographers have suggested classifying towns according to their activities or functions. This is not very satisfactory because towns sometimes change their function: their original function may have been superseded by another; and towns seldom have only one function: most large towns are multi-functional.

The following suggested classification gives a simple grouping into eight categories and largely follows Aurousseau, who proposed it in 1921.

(a) *Administration*. These may be national or local centres:
　(i) Capitals, e.g. Washington, Canberra.
　(ii) Regional administrative centres, e.g. Wakefield.

(b) *Defence*. These may be sub-divided into three types:
　(i) Fortress towns, e.g. Ludlow, Caernarvon, Frederica (Denmark), Delhi.
　(ii) Garrison towns, e.g. Aldershot.
　(iii) Naval bases, e.g. Plymouth, Toulon.

(c) *Culture*. Interpreting culture widely, there are two kinds of centres:
　(i) Religious, e.g. Canterbury, Lourdes, Jerusalem.
　(ii) Seats of learning, e.g. Cambridge, Heidelberg.

(d) *Production*. This group includes commercial as well as industrial towns:
　(i) Extractive centres or mining towns, e.g. Genk (Belgium), Strassfurt, Khouribga (Morocco).
　(ii) Manufacturing centres, e.g. Birmingham, Essen, Detroit.
　(iii) Trading and financial centres, e.g. Dusseldorf, Zurich, Winnipeg.

(e) *Transport*. Two examples of this type of town are as follows:
　(i) Ports, e.g. Liverpool, Le Havre, Rotterdam.
　(ii) Railway centres, e.g. Crewe, Swindon, Chicago.

(f) *Recreation*. These are tourist and health centres and may be sub-divided into three:
　(i) Spas and watering places, e.g. Tunbridge Wells, Baden-Baden.
　(ii) Seaside resorts, e.g. Blackpool, Cannes, Palm Beach.
　(iii) Sports centres, e.g. St. Moritz, Banff.

(g) *Residential or dormitory towns*. The main function of many older settlements is now to serve nearby larger towns as residential areas, e.g. Cheadle Hulme for Manchester, Haywards Heath for London, Vallingby for Stockholm.

(*h*) *New towns.* There is an increasing number of towns which have been, or are being, deliberately created by government planning. Early examples of new towns were Welwyn Garden City and Letchworth; Washington New Town and Peterlee are more recent examples. It will be useful to elaborate further upon this category of towns in **18–21** below.

NEW TOWNS

18. The idea of new towns. First, it should be emphasised that the idea of building completely new towns is not new: it goes back many centuries. The concept of the "new town" means that a town, instead of growing naturally out of a village as a result of expanding population, trade and increased nodality, is deliberately planned and planted. There seem to be definite historical phases when the creation of new towns became popular. In England, the thirteenth and fourteenth centuries were an age of pronounced civic activity, and since the Second World War there has been another surge of activity.

19. Reasons for building new towns. Whatever the reasons were in the past (they seem to have been largely military and commercial), at present three principal reasons explain the creation of new towns:

(*a*) The population is increasing, particularly in south-east England.

(*b*) The overspill population from some of the larger cities, especially London, needs housing: this is an attempt to restrict the growth of existing towns.

(*c*) The provision of new growth points in derelict or declining areas also gives better housing and improved amenities to the people living in those areas.

20. Pre-war developments. An early nineteenth century attempt at new town planning was Saltaire, near Bradford, undertaken by Sir Titus Salt, a wool textile manufacturer. But one of the first of the modern town planners was Ebenezer Howard, who designed Letchworth at the beginning of the present century.

After the First World War there was a great housing shortage and many local government authorities attempted to solve the housing problem by building huge housing estates. For example, the London County Council built St. Helier in Surrey (outside the L.C.C.'s jurisdiction), converting a large area of farmland

FIG. 20 *New towns in Britain.*

Since the war many new towns have been built in rural areas to accommodate "overspill" populations from large towns. Though built quickly, they have been carefully planned. In addition to the new towns, many other small towns have been expanded; and others have been planned to undergo development and enlargement.

into a suburban area. St. Helier was purely a section of suburbia with no definite centre, no factories and no main shopping area. Manchester built an enormous estate at Wythenshawe which now houses 90,000 people. Welwyn was another of these experiments before the Second World War.

21. The New Towns Act 1946. The purpose of this Act was to create a number of completely new towns. The aim was to build self-contained civic units with civic centres and industrial estates which would provide employment for the town's inhabitants.

For example, Harlow New Town in Essex, 37 km from London, was built under the control of the Harlow Development Corporation. The new town was built west of the A11 and the old village of Harlow; it was planned with five residential areas, each having its own amenities, grouped round the town centre which was arranged as a precinct from which traffic was largely excluded. The various residential areas are separated from one another by belts of open country which extend right into the centre of the town. Two industrial estates, the Pinnacles and Templefields Estates, are located on the northern fringe of the town near the railway and near main highways. Harlow is a clean, spacious, well-planned town and has been a successful experiment. Its target population is 130,000; at present it has a population of about 80,000 (*see* also Table III and Fig. 20).

MILLION CITIES

22. The growth of big cities. Historically there have always been great cities with large populations, e.g. Rome, Angkor, Hangchou, but the large city is very much a creation of the last hundred years and especially the last thirty years.

The number of towns possessing over one million inhabitants has grown rapidly: in pre-war days there were about 25, just after the war about 36, in 1953 about 64, in 1960, 83 and in 1975 over 100. These million cities, or millionaire cities as they are sometimes alternatively called, are an urban phenomenon of every continent but they are least prominent in Africa and Australia. At present there are some two dozen cities with populations in excess of three million inhabitants.

23. Distribution. The general distribution of these million cities is of considerable interest and a study of a world map showing their distribution (*see* Fig. 21) reveals these facts:

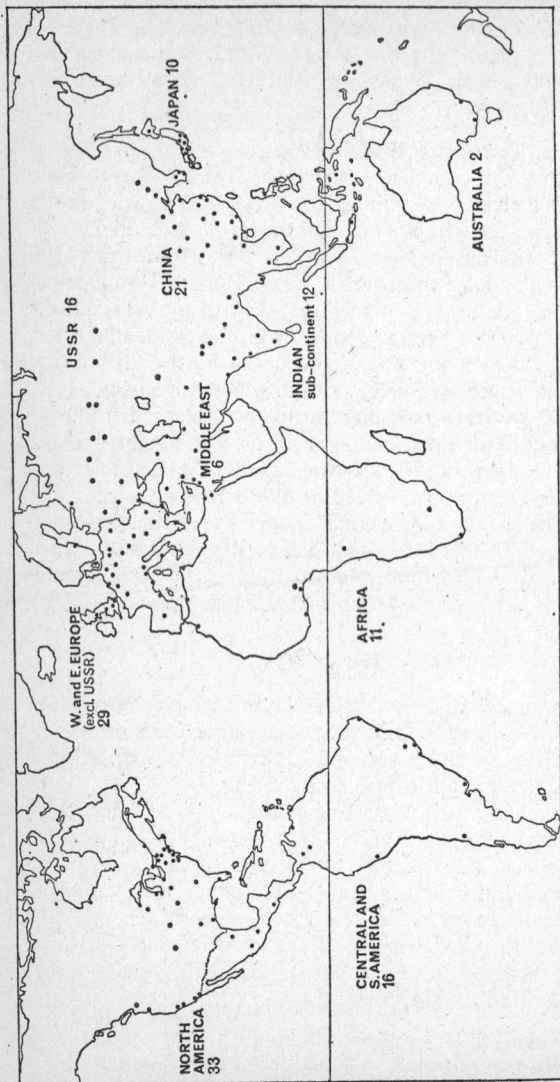

FIG. 21 *Distribution of million cities.*

In 1900 there was only a handful of cities with a population exceeding 1 million; in 1965 there were over 80, in 1975 over 100. At least 24 of these have populations of over 3 millions. Most of these million cities lie in temperate latitudes and in regions of high population density.

Map labels:

NORTH AMERICA 33

W. and E. EUROPE (excl USSR) 29

USSR 16

MIDDLE EAST 6

CHINA 21

JAPAN 10

INDIAN sub-continent 12

AFRICA 11.

CENTRAL AND S. AMERICA 16

AUSTRALIA 2

(a) Relatively few lie within the tropics.

(b) Relatively few are in the southern hemisphere.

(c) A large proportion have coastal locations.

(d) The majority occur in the earth's most densely peopled areas.

24. Locations. A careful study of their locations shows the following:

(a) The overwhelming majority lie in lowland areas, although there are some exceptions, e.g. Madrid, Mexico City, Bogota.

(b) A very high proportion are sea-, lakeside or river-ports; e.g. Buenos Aires, Chicago, Hang-chou.

(c) Many possess exceptional advantages of site, e.g. New York, Paris, Vienna, Istanbul, Cairo.

The earlier million cities were predominantly either old-established state capitals or advantageously situated ports; they had, as D. C. Money has said, "some outstanding attributes of location, or history" which enabled them to dominate other towns in the same general area. The more recent million cities, however, show a high proportion of basically industrial towns; this is true of many of the new Soviet, Chinese and Indian million cities, e.g. Harbin, Taiyuan, Ahmadabad, Kanpur, Lahore.

25. Reasons for growth. A question which naturally arises is: why have so many cities in recent decades grown to be million cities? It is difficult to be precise or sure as to the reasons but, without doubt, some of the following conditions have helped:

(a) In many regions the growth in industrialisation has promoted the migration of the excess rural labour force into the towns.

(b) In some countries, such as Japan, shortage of arable land has compelled people to seek alternative employment in the towns. (Note that this, a compelling factor, differs from (a), where industry, with perhaps higher wages, is an attracting factor.)

(c) In some areas the mechanisation of agriculture has led to rural unemployment and the peasants have thus been compelled to seek work in the towns.

(d) As people become more educated they desire "better jobs" and the opportunities are practically confined to the urban areas.

TABLE III. NEW TOWNS IN GREAT BRITAIN

Towns under Development Corporation	Date designated	Approx. pop. at designation	Original pop. target	Latest revised target	Main purpose for which town was designated
Stevenage	Nov. 1946	6,700	60,000	105,000	London overspill
Crawley	Jan. 1947	9,100	56,000	85,000	London overspill
Hemel Hempstead	Feb. 1947	21,000	60,000	80,000	London overspill
Harlow	Mar. 1947	4,500	80,000	130,000	London overspill
Aycliffe	Apr. 1947	60	20,000	45,000	Housing dispersed workers on industrial estate in Durham
East Kilbride	May 1947	2,500	50,000	100,000	Glasgow overspill
Peterlee	Mar. 1948	200	30,000	—	Rehousing Durham miners
Hatfield	May 1948	8,500	25,000	30,000	London overspill
Welwyn Garden City	May 1948	18,500	50,000	—	London overspill
Glenrothes	June 1948	1,100	30,000	75,000	Housing workers in new colliery in Fifeshire. Later Glasgow overspill
Basildon	Jan. 1949	25,000	100,000	103,000	London overspill
Bracknell	June 1949	5,000	25,000	60,000	London overspill
Cwmbran	Nov. 1949	12,000.	35,000	55,000	Better housing for workers in South Wales
Corby	Apr. 1950	15,700	40,000	80,000	Housing steel workers in East Midlands
Cumbernauld	Dec. 1955	3,000	50,000	100,000	Glasgow overspill
Skelmersdale	Oct. 1961	10,000	80,000	—	Merseyside overspill
Livingston	Apr. 1962	2,000	70,000	100,000	Glasgow overspill
Dawley, later Telford	Jan. 1963 & Dec. 1968	21,000 & 70,000	55,000 —	25,000	W. Midlands overspill

	Date				
Redditch	Apr. 1964	31,500	70,000	90,000	W. Midlands overspill
Runcorn	Apr. 1964	28,500	70,000	100,000	Merseyside overspill
Washington	July 1964	20,000	60,000	80,000	Tyneside & Sunderland overspill
Craigavon	July 1965	40,000	100,000	180,000	Belfast overspill and economic development of N. Ireland
Antrim	July 1966	3,000	30,000	—	Economic development of N. Ireland
Irvine	Nov. 1966	30,000	70,000	120,000	Glasgow overspill
Milton Keynes	Jan. 1967	40,000	250,000	—	Immigration into S.E. England
Peterborough	Aug. 1967	81,000	175,000	200,000	London overspill and immigration into S.E. England
Ballymena	Aug. 1967	18,000	70,000	—	As Antrim
Newtown	Dec. 1967	5,500	13,000	—	Economic growth of C. Wales
Northampton	Feb. 1968	131,000	220,000	> 250,000	As Peterborough
Warrington	Apr. 1968	75,000	> 200,000	225,000	Manchester overspill
Londonderry	Feb. 1969	82,000	94,000	100,000	Economic and social development of N. Ireland
C. Lancashire	Mar. 1970	235,000	400,000	430,000	Economic development and social improvement of C. Lancs.
Towns under County Council					
Killingworth	1959	310	17,000	20,000	Tyneside overspill
Cramlington	1963	7,700	48,000	62,000	Tyneside overspill

(e) The social amenities and attractions of town life have led to a marked drift in many areas into the larger urban centres.

26. Social implications. In conclusion, three points of considerable social importance may be noted:

(a) The exaggerated growth of many metropolitan centres is presenting acute problems of housing, sanitation, water supply, etc. This is particularly the case in India and Latin America; in the latter area many of the capital cities, e.g. Buenos Aires, Montevideo, Santiago, Caracas, contain an undue proportion of the population of their respective countries.

(b) In the less well-developed countries there has been an influx of native rural people into the towns, with attendant overcrowding and the growth of sordid slums. Around many of the great cities in South America, for instance, unorganised, unplanned shanty-towns, known as *favellas* in Brazil, have suddenly appeared. The same is true of Tunis, Johannesburg and Calcutta.

(c) There is commonly acute urban unemployment in the big cities of the developing world. Those people unable to find work in farming or rural industries often leave for the towns hoping to find a job there. But, in almost all cases, there are not enough jobs for the many who come. There is a great need in the towns for more small-scale industries which employ many people. Only by developing labour-intensive industries and expanding service industries can the high proportion of unemployed hope to be reduced.

27. Megalopolis. Accelerating urban growth during the twentieth century, aided by improvements in transport, ultimately led in some areas to the coalescence of neighbouring towns to form large continuously built-up urban agglomerations or conurbations. If this process of urban accretion should continue it is feasible that some of the conurbations could coalesce to create continuously urbanised regions. Professor Jean Gottmann has suggested that the entire north-eastern seaboard of the United States, extending from New Hampshire to Maryland, may be visualised as a potential single urban complex. To such a vast urban region, he has given the name megalopolis. Already this north-eastern seaboard region has a population of 35 millions and dwarfs even the largest conurbation. There are other areas in the world where a megalopolis could develop; indeed, it has been suggested that by the early part of the twenty-first century the whole of south-eastern England could have developed into a vast urban region.

PROGRESS TEST 8

1. In what different ways may a town be defined? Which do you think is the most satisfactory? **(1)**

2. Towns are described as "dynamic organisms". What is meant by this? **(3)**

3. Which factors or conditions may be said to give a town a personality? **(4)**

4. Give appropriate examples of: a newborn baby town, a quiet studious town, a fat millionaire town, a dying town. **(4)**

5. What is meant by the term "the urban revolution"? When did it take place and where? **(5)**

6. Trade has been said to be the *raison d'être* of the town. Attempt to justify this statement. **(7)**

7. Medieval towns owed their origin to three main causes. What were these causes? Give examples of each. **(10)**

8. Which causes have been responsible for the decline or destruction of towns? **(11)**

9. Explain the meaning of the following: ghost town, *bastide* town, planned town. **(11, 13)**

10. What is meant by the phrase "town structure"? **(14)**

11. Attempt a classification of towns according to their origin. **(16)**

12. Give examples of the following: a gap town, a confluence town, a bridge-point town, a defensive town, a spa town. **(16)**

13. Why has it been found necessary or desirable to build new towns in Britain? **(18, 19)**

14. What is meant by the term "million city"? Approximately how many are there at the present day? **(22)**

15. What reasons lie behind the growth of million cities? **(25)**

CHAPTER IX

Political Geography

STATE, NATION AND NATIONALISM

1. The scope of political geography. Political geography is the field of study where politics and geography overlap and mutually influence each other. Let us quote two definitions of its scope.

According to N. J. G. Pounds:

"Political geography is concerned with politically organised areas, their resources and extent, and the reasons for the particular geographical forms they assume."

F. J. Monkhouse defines political geography as:

"The study of states, their frontiers and boundaries, their inter- and global relations, their contacts and their groupings; the variation of political phenomena from place to place, considered in relation with other features of the earth as the home of Man."

In this chapter we shall deal with various aspects of the study.

    (*a*) The territory of the state.
    (*b*) The resources of the state.
    (*c*) State frontiers and boundaries.
    (*d*) Communications.
    (*e*) State capitals.
    (*f*) Political groupings.
    (*g*) Geopolitical ideas.

2. Countries. A glance at a political map of Europe, for example, shows that the earth's land surface is divided into political units or countries, some very large, some of moderate size, some small and some diminutive (*see* Fig. 22). The question arises: why should states vary so much in size; why did some become so big, yet others remain so small?

If, for example, we look at Europe we find that it is fragmented into a large number of moderately sized and small sized states. Why are there, in such a relatively small territorial area, so many

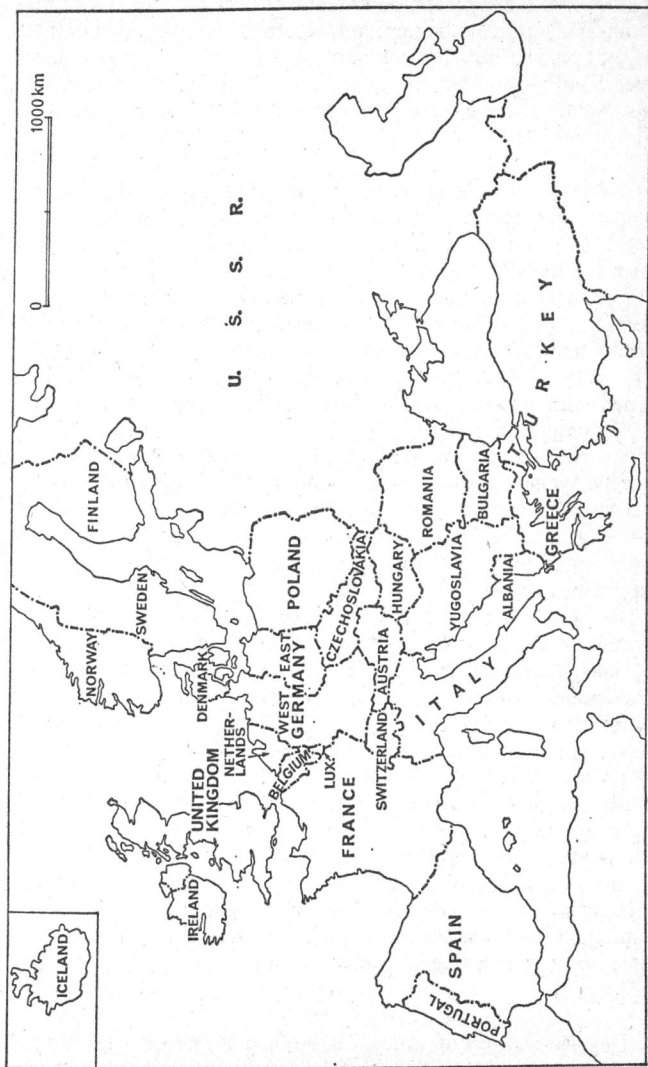

FIG. 22 Political map of Europe.

separate and independent states? The answer is to be found in the interplay of many factors; of race, language, religion, nationality, physical configuration, topography and history. On the other hand, why does the Soviet Union cover approximately one-seventh of the earth's land surface? All these questions are of great interest.

3. The territory of the state.

Much confusion exists in the minds of many people as to what constitutes nationality, and the difference between state and nation. Let us begin by attempting to separate and define them.

A primary condition for the formation of any state is the possession of a certain expanse of territory, known as the territorial domain. A state, therefore, may be defined as the political organisation of a given territory. Essentially the state is concerned with political organisation and territorial domain, and implies unfettered sovereignty.

Two well-known French geographers, Jean Brunhes and Camille Vallaux, wrote that the political association of the state was born of the necessity of collective security, and that this necessity appeared only when people, occupying territory and exploiting it to meet their needs, felt that they had a common patrimony to defend.

The nomadic way of life has never been conducive to the formation of states, for, although nomadic groups frequently recognise particular areas as being theirs, their unsettled life has made political organisation of an era difficult and, in the face of a threat, they usually move on instead of standing their ground. Moreover, nomadic groups are often too few in numbers and too widely dispersed to occupy territory effectively. This leads on to another of the axioms of state formation: there must be a certain minimum density of population. This helps to explain why such groups as the Eskimos and the Lapps never developed states.

The presence of other groups of people occupying adjacent territory who were or were likely to be rivals or enemies has been, historically, a potent factor helping to cement groups. External threat, real or imagined, emphasised and engendered the need for security.

4. The rise and fall of states.

To summarise, we may say that, historically, the conditions requisite for the formation of a state were mainly three:

(a) The permanent occupation of a territory.
(b) A minimum density of population.
(c) The contiguity of rival or opposing peoples.

Even when these conditions exist, the development of the state does not always proceed with the same degree of success. In some places a succession of states emerge, rise and fall, the new states arising out of the old. In other places one finds states developing but lacking sufficient strength and internal cohesion to maintain themselves and coming under the influence of, or being incorporated by, their more vigorous neighbours.

If there is one thing that is certain it is that no state lasts for ever, no matter how powerful it may be at a given point in historical time. It is sometimes alleged that China provides the exception to the rule: that China has existed as a state for some 4,000 years. This is untrue: while China has exhibited an astonishingly unbroken cultural continuity, the Chinese state has fallen asunder many times, a new China rising phoenix-like out of the ashes of the old state.

5. Nationality. The term nation is not an alternative expression for state. The distinction between them is perhaps best made clear by saying that the state refers to territory, the nation to people.

Man is by nature, by instinct, a gregarious creature. Group life, no matter how simple or primitive, has some form of social organisation. Into this category of social groups fall the family, the clan, the tribe and the nation.

The nation as a group is not easy to define; in fact, nationality is a somewhat intangible concept; but, though often rather nebulous, the national sentiment is very real and frequently strongly expressed and held.

The nation can, perhaps, be adequately defined as a group of people who feel themselves bound together through personal ties and who possess a coherence and solidarity which has grown up through the following influences:

(a) By following a common way of life,
(b) By sharing common experiences.
(c) By possessing common cultural traits.
(d) By inheriting a common tradition.

This idea of nationalism—the expression of the sentiments of a group—again is not novel: it is virtually as old as man himself. What is new about it is its relationship to the state.

6. The criteria of nationality. There are many reasons for the creation and sustainment of national feeling:

(*a*) *Race.* Race may produce national feeling but, as we have already pointed out, there are practically no pure racial groups (*see* II). Where race has played a part, it has usually been invoked spuriously. For example, the Nazis spoke of the superiority of the "German race" when, in truth, no German race or even ethnic group existed; the Germans are mostly a mixture of Alpine and Nordic stocks, sub-groups of the Caucasian family.

(*b*) *Language.* Language is an important element fashioning national groups. As W. A. Gauld wrote: "As a bond, language may be the principal means of creating common sentiments and traditions which go to make a grouping such as nationality. It contributes to cultural unity by facilitating the expression of common experience and achievement. It may go further and stimulate political unity by the fostering of organised institutions. Far more than ethnic unity, language may encourage political and social affinity" (*Man, Nature and Time*). Nationalism in nineteenth- and twentieth-century Europe was very largely geared to, and was an expression of, language.

NOTE: Linguistic unity is not a prerequisite of national unity as the quadrilingual country of Switzerland demonstrates, but it is certain that a common language acts as a potent cementing factor in a social group.

(*c*) *Religion.* Historically, religion was important as a formative influence in politics. The Protestant challenge to the Roman Catholic Church in the sixteenth century helped nationalism since it involved a repudiation of foreign influences. In the Protestant countries the Churches became largely nationalised, e.g. the Anglican Church. Today religion plays a less significant role in politics, although it is still important in Ireland, and the division of India in 1947 was fundamentally related to the religious conflict between Muhammadanism and Hinduism (compare also the creation of Bangladesh in the 1970s).

(*d*) *Geographical setting.* Political evolution, and the growth of nationalism in western Europe especially, has been greatly assisted by the geographical setting. The insular character of Britain, providing a measure of isolation and security, is an obvious geographical element encouraging independent development. The large, fertile and rich Paris Basin provided an obvious focal point for the growth of the French state. The mountain-framed "Bohemian diamond" provided a fairly clear-cut entity for the growth of Czech nationality.

(e) *Common enemy.* Rivalry, dislike and hatred between two groups will stimulate national feeling. The security of the group will weld its peoples firmly together and create a unity of national feeling. This is amply demonstrated by the antagonisms which have existed between France and Germany, Finland and Russia, the Muslims and the Hindus, and the Arabs and the Jews.

7. The nation-state. In referring to the countries of western Europe, it is customary to call them "nation-states". Such a term implies not only that the countries occupy territorial areas but that those areas are also occupied by national groups. For example, in France the nation is virtually coterminous with the state or geographical area of France. This condition applies generally to most of the countries of western Europe (the notable exception here is Belgium which comprises two quite distinct national groups: the Flemings and the Walloons).

The fundamental difference between state and nation can perhaps best be illustrated by reference to the former Austro-Hungarian Empire and the present Soviet Union. The Austro-Hungarian Empire comprised a single political unit or state but it was made up of a large number of distinct national groups—Austrians, Hungarians, Czechs, Slovaks, Croats, Italians, etc. Similarly, the Soviet Union is composed of numerous national groups, e.g. Russians, Ukrainians, Karelians, Georgians, etc., but it forms a single state.

8. The importance of nationality. The idea of nationality emerged strongly after the French Revolution and nationalism has become a potent factor in world politics. People respond to national feeling. It is very significant that during the Second World War when the Germans invaded the Soviet Union, Stalin appealed to the Soviet peoples to save not Soviet communism but Mother Russia. While the countries of western Europe have to a large extent outgrown nationalism and have been thinking in terms of a supra-national authority, many countries in the rest of the world have only just become conscious of nationalism; the emergent countries of Africa, the Middle East and Latin America are all going through the early active and vociferous stages of nationalism.

SHAPE, SIZE AND SITUATION

9. Primary elements. The late American geographer S. van Valkenburg wrote: "The three primary elements in a political-geographical evolution are location, size and shape. The political geographer deals with the problems of nations, and when he considers any individual country or group of countries his first questions are: "Where is it?", "How large is it?" and "What is its shape?" (*Elements of Political Geography*).

Each of these elements is of great importance and they "play an important role when associated with various other elements in a political-geographical evaluation". Let us first, therefore discuss these three concepts. It will be useful to have a map showing the political pattern of the world for reference.

10. Geographical location. The location of a place is a geographical factor of much significance. Location implies two things: the first of these is constant, the second is alterable:

(*a*) *Absolute location* as defined by lines of latitude and longitude: this is mathematically determined and unalterable.

(*b*) *Relative position*, that is, position in relation to other areas, such as land areas or bodies of water. Relative position also includes accessibility, which is of particular importance. For example, an area, e.g. a country or part of a country, may have its development retarded because it is remote, isolated and inaccessible—think, for instance, of Bolivia, Tibet and the northern parts of Canada or the Soviet Union, or even Australia until about one hundred years ago.

11. Examples of changes in relative position and accessibility. The relative position and accessibility of a place are not constant; they may change as a result of new communication developments. Think of the changes which have occurred as a result of the elimination of obstacles and the introduction of new means of transport. Three illustrations may be given to show how the relative position and accessibility of areas have changed:

(*a*) In Roman times the effective centre of the known world (the *oecumene* as it was called) was the Mediterranean and in those days Britain lay remote on the edge of that world. With the discovery of the Americas, around 1500, the commercial centre of gravity moved to western Europe and Britain then came to lie in the midst of the most important portion of the earth's surface —a position which she continues to hold and which brings enormous advantages with it.

(b) The cutting of the Suez and Panama Canals brought great benefits to many countries: not only was travelling time, and therefore cost of carriage, greatly reduced between certain countries but, as a result of these ship canal developments, economic production and trade were greatly stimulated, e.g. along the Pacific coasts of North and South America, regions which, hitherto, had been rather remote, isolated and economically backward.

(c) The advent of the aeroplane, a completely new means of communication which soon developed speeds undreamed of in land-bound and water-surface transport, shortened drastically the travelling time over long distances and also made accessible regions which formerly had been extremely difficult of access. There are, even today, many places which are unconnected by either rail or road but which are served by aircraft; e.g. many of the Soviet arctic settlements fall into this category.

12. Size. Size is important, for without sufficient size no country can ever rank as a leading world power. It is no accident that the two great world powers of the present day, the United States and the Soviet Union, are territorially large states. On the other hand, large size does not inevitably imply greatness, strength and power. For instance, both Canada and Australia rank among the largest states in the world but neither could lay claim to being a major world power.

States may be classified according to size roughly as follows:

(a) *Very large:* Soviet Union, United States, Canada, Brazil, Australia, China.

(b) *Large:* South Africa (inc. Namibia), Argentina, Mexico, India, Indonesia, Saudi Arabia, Sudan, Zaïre.

(c) *Medium size:* France, Egypt, Pakistan, Peru, Venezuela, Ethiopia.

(d) *Small:* Belgium, Netherlands, Lebanon, Dominican Republic, Liberia, Taiwan.

(e) *Diminutive:* Luxembourg, Andorra, Liechtenstein, Monaco, San Marino.

NOTE: Some countries have attained an importance in the world which is altogether out of proportion to their size, e.g. Switzerland and the Netherlands. Though small countries may attain considerable significance, perhaps because of their strategic location, their commercial expertise or their highly developed culture, they are, in the final analysis, severely handicapped by their size.

13. The importance of size. Spatial extent is important for these reasons:

(*a*) Larger areas usually imply enlarged resources and greater sectional diversity of agriculture. For example, the gigantic size of the Soviet Union means that she has a share of practically every natural resource on the face of the earth while the United States, situated in mid-latitudes, is favoured by a wide range of climatic conditions.

(*b*) Size gives a greater measure of national security. During the Second World War the Soviet Union was able to organise defence in depth: she had plenty of room to manoeuvre and retreat. Today, with the threat of atomic warfare, sheer space is a great advantage; while a small area could be easily saturated with missiles, a large area could not be entirely devastated.

14. Shape. The shape of a country may be advantageous or disadvantageous. Clearly, the more compact the shape the better. Theoretically, a circle is the ideal shape because it gives the maximum degree of compactness, and the boundary length in relation to area is at a minimum.

No state is, of course, circular in shape but some approximate fairly closely to this form; e.g. France is roughly hexagonal in shape, and so are Spain, Poland, Romania, Cambodia. Countries which are roughly square include Egypt, Libya, Nigeria and Bulgaria.

Some countries are elongated. Such long, narrow shapes bring difficulties of national cohesion, though this may be offset by providing increased climatic variation; Chile is an example. Other elongated countries are Norway, Italy and Japan.

There are other shapes, especially those where the state is separated into units. Van Valkenburg (*Elements of Political Geography*, Pitman), suggested a threefold division (*see* Fig. 23):

(*a*) *A broken shape*, where one or more sections are disconnected from the main territorial area, e.g. the United States if one includes Alaska, pre-war Germany.

(*b*) *A fragmented shape*, where there are numerous fragmented

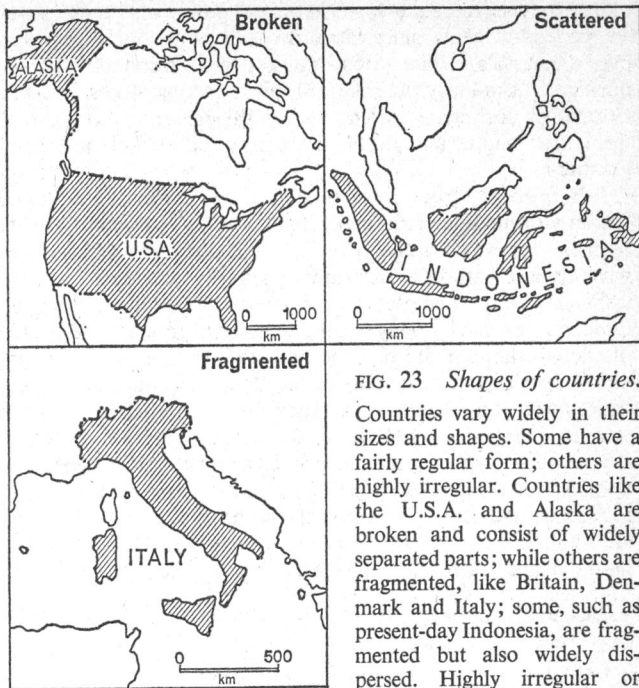

FIG. 23 *Shapes of countries.*

Countries vary widely in their sizes and shapes. Some have a fairly regular form; others are highly irregular. Countries like the U.S.A. and Alaska are broken and consist of widely separated parts; while others are fragmented, like Britain, Denmark and Italy; some, such as present-day Indonesia, are fragmented but also widely dispersed. Highly irregular or awkward shapes and fragmentation and dispersion usually bring problems to national unity.

parts but where there is usually a main territorial mass as, for example, in Italy or even the United Kingdom (Britain, Northern Ireland, all the offshore islands, the Isle of Man, the Scillies, Channel Islands, etc.).

(*c*) *A scattered shape*, best exemplified today in the case of Indonesia but represented historically by the Republic of Venice.

RESOURCES AND COMMUNICATIONS

15. The resource base. These days a highly developed industrial economy is a prerequisite of great power status. Any state lacking adequate power and mineral resources is at a disadvantage. As was pointed out earlier in **13**, the larger the territorial area the

more likely is the state to have varied power and mineral resource endowments since these are not equitably distributed
over the earth's surface. Most states attempt to industrialise for
prestige and military purposes. Britain's past greatness, pre-war
Germany's formidable strength, and the present-day might of
the United States and the Soviet Union reflect their industrial
greatness.

It is probably true to say that no state, not even the Soviet
Union with its huge land area, is able to live as an autarky, i.e.
by using only such native resources as it possesses. Often states
have tried to attain a measure of autarky and national self-
sufficiency was attempted in the inter-war period, notably by
Germany. Even Germany would never have reached the degree
of success that she did had she not achieved a measure of economic domination over many of the Danubian countries. A pro-
tracted war began to show how the countries with an inadequate
resource base, especially Italy and Japan, were very vulnerable.

Many countries, Britain in particular, are largely dependent
upon external supplies of power, industrial raw materials and
foodstuffs. Access to these and the security of communications
are vital. In both World Wars the submarine menace was perhaps the most serious threat to Britain's existence.

16. Communications. The role of communications in politics
falls under two headings:

    (*a*) Internal communications (*see* **17** below).

    (*b*) External communications (*see* **18, 19** below).

17. Internal communications. One of the fundamental problems
of the state is to ensure the cohesion of the various parts of which
it is constituted. This cohesion and unity can be secured by lines
of communication. So accustomed are we nowadays to think of
communications as serving economic ends that we are apt to
forget their political purpose. Historically, the primary purpose
of communications was political and military.

    (*a*) The great highway built by Darius the Great, probably the
first major road built by man, formed the great central artery of
the Persian Empire.

    (*b*) The great network of roads constructed by the Romans,
which were long, straight and strategically located, were military
highways in essence, binding together the component parts of the
widespread Empire.

(*c*) The extensive network of roads in France, first laid down by Napoleon, was built and maintained, fundamentally, for strategic purposes.

(*d*) The magnificent *Autobahnen* or arterial highways built by the Nazis in Germany during the inter-war period were aligned mainly east–west and were constructed with one primary aim in view—latent military needs.

NOTE: It is interesting to consider how many successful trade routes owe their origin to military considerations.

(*i*) The Roman roads are still followed by many of the great highways of Europe.

(*ii*) General Wade's military roads in Scotland subsequently became commercial highways.

(*iii*) The Trans-Siberian and Trans-Caspian railways in the Soviet Union were military in origin.

(*iv*) The Alaska (Alcan) Highway, built during the Second World War as a military highway, now functions as a commercial routeway.

18. External communications. Two groups of countries in particular are especially concerned with external communications:

(*a*) Off-shore island states, e.g. Great Britain, Japan, Sri Lanka.

(*b*) Land-locked states, e.g. Switzerland, Afganistan, Bolivia, Rhodesia.

The one requires freedom of the sea approaches; the other requires overland access to the sea. One need not emphasise the dilemma of such states if they are cut off from contact with the outside world through the hostility of neighbours.

19. Control of external communications by states. Certain countries have great strategic significance because of their command of important routeways. In some cases they have taken advantage of this to deny to other countries freedom of access to these means of communication:

(*a*) *Countries that control routeways:*

(*i*) Denmark controls the Öresund, the principal sea entrance to the Baltic.

(*ii*) Turkey controls the Bosphorus–Dardanelles entrance and exit of the Black Sea.

(*iii*) Egypt controls the Suez Canal, the short-cut to the East.

(*iv*) England can effectively control the English Channel.

(*b*) *Countries that are denied the use of routeways:*

(*i*) The roads linking Tibet and India have been cut by China.

(*ii*) On occasions Bolivia has had her rail links severed.

Another important aspect of communications relates to air space and air routes. The advent of aircraft revolutionised warfare. Fear of aircraft, not only for bombing purposes but for spying, has caused many states in the past to prohibit flying over their territories. Aircraft have, to a very considerable extent, rendered navies obsolete.

FRONTIERS AND BOUNDARIES

20. Frontier and boundary. Some confusion exists over the terms "frontier" and "boundary"; often they are used inter-changeably to mean the same thing but this is incorrect for they refer to two different things: a frontier is a zone, a boundary is a line.

A frontier is a zone or belt of territory, a no-man's land, which separates one group of people from another. It may be referred to, sometimes, as a border, e.g. the Scottish border country (the Southern Uplands), or as a march, e.g. the Welsh marches.

A boundary is a line, sometimes demarcated on the ground; it is specific. The term frontier line might be taken to be synonymous with boundary but not frontier.

21. European state boundaries. A study of European state boundaries shows their development through a series of stages from the frontier zone to the definitive boundary line. First, there was a wide and little-inhabited zone between two groups of people; next, the more advanced group built a chain of fortifications as a protection; then the vacant zone disappeared but the fortified points remained; finally a wholly continuous frontier line was laid down, either by imposition or by mutual agreement.

The Pyrenees, situated between France and Spain, form a rugged, mountainous, sparsely peopled frontier zone. Since the Treaty of the Pyrenees in 1659 between the two countries there has been no tension in this region and the Pyrenees have served as a good border. Often, however, there is active tension in the frontier zone and, accordingly, any boundary line which is established is apt to change frequently to meet the changing balance of forces, e.g. the boundary between France and Germany.

22. Functions of frontiers and boundaries. Their function is to delimit the area of sovereignty of the state. Within those bounds the people choose the way of life they prefer and, in democratic societies, have the freedom to choose the form of government they want.

The original function of the frontier as a defence mechanism hardly applies at the present day, for modern modes of warfare have rendered the frontier's defensive function obsolete.

Frontiers should not prohibit intercourse and the free exchange of ideas and goods but sometimes they do as in the case of the boundary between West and East Germany and that between Gibraltar and Spain which is closed.

23. Natural frontiers. Apart from the sea, there are no natural frontiers in the absolute sense of the term; history demonstrates that neither mountains nor rivers nor any "natural" barrier serve as insurmountable obstacles to invading armies (*see* below, **24–28**). Nevertheless, we do recognise certain types of frontiers as "natural".

24. Sea. The open sea offers the best natural frontier. Although seas can be crossed, and have been crossed by invading forces many times in history, they act as a good defensive shield, particularly if the country has an efficient navy. Although the English Channel is a mere 33 km across at its narrowest point, Britain has not been invaded since 1066. The greatest drawback to the sea frontier is that it may lead to the development of an insular outlook; for example, Britain has seldom been interested in continental affairs unless her security has been at stake.

25. Mountains. Boundaries usually follow the crest or main watershed. The altitude, ruggedness and steep slopes of mountain areas perform a separative function and, in general, facilitate defence. Movement is difficult and is usually channelled through passes, e.g. the Alps. Mountains, even the highest, do not provide an insuperable barrier and the Himalayan system which has provided a most effective barrier has not prevented wave after wave of invaders penetrating into India, mostly through the Khyber Pass. However, highlands have formed fairly effective zones of separation: think of the Southern Uplands of

Scotland, the Scandinavian Highlands, the Carpathians, and the southern Andes between Argentina and Chile.

26. Deserts. Desert areas, because of their lack of water and the difficulties of movement across them, have formed empty zones and served as barriers. The Sahara, for example, has provided *the* great physical and human dividing line in Africa; it is often said that Africa proper begins south of the Sahara. In the early days of human history Egypt was protected by the surrounding desert and this permitted the undisturbed development of an early civilisation in the lower Nile Valley. In much the same way the Gobi Desert gave China a measure of protection along her north-western frontier.

27. Rivers. Rivers have the advantage of forming readily discernible and well-marked lines but their efficiency as boundaries is reduced for two main reasons:

(*a*) Rivers are apt to change their courses and hence problems may arise as to the line of the original boundary.

(*b*) Rivers and their valleys are inclined to attract and cause concentration of settlement and so are not particularly successful as barriers between peoples.

Even so, rivers frequently serve as boundaries, e.g. sections of the Rhine, Danube, St. Lawrence, Rio Grande, Paraná–Paraguay.

28. Forests and marshes. Thick forests and swamplands are often difficult to penetrate and cross; this was more particularly true in earlier times—and so historically they have often served as frontier zones. The great Mohakontara forest effectively separated peninsular India from the Indo-Gangetic Plain while forests and fenlands cut off East Anglia from the rest of the English Plain. The forests of the Baltic Heights helped to keep the Letts and Lithuanians free from the Russians. Perhaps the best example of a marsh barrier is the Pripet Marshes which separated Poland and Russia for many years. Few marshland barriers exist today because of the need for agricultural land.

29. Artificial boundaries. These are of varying kinds but they are all essentially "artificial" in that they are man-made. They may be divided into five types:

(*a*) *Defensive walls.* In earlier times man constructed "walls" which were fortified, e.g. Offa's Dyke along the Welsh border, Hadrian's Wall and the Great Wall of China. A modern example of this kind of boundary is the Berlin Wall.

(b) *Demilitarised zones.* Sometimes "empty" or demilitarised zones function as boundaries. The neutral territory between Spain and Gibraltar falls into this category, similarly the demilitarised zone between North and South Korea although here the 38th parallel is the treaty line.

(c) *Latitude and longitude.* A striking feature of the political boundaries in North America, Australia and parts of Africa is the use of mathematical lines of reference, e.g. the 49° N. between the United States and Canada. Parallels and meridians are often used in areas where occupation and settlement have proceeded more rapidly than accurate survey.

(d) *Geometrical lines.* Sometimes boundary lines are drawn obliquely across an area from point to point: examples are the southern boundary between Kenya and Tanzania, the boundary between Algeria and Mali. These lines, like lines of latitude and longitude, ignore natural features completely.

(e) *Lines of mutual agreement.* Sometimes the boundary between one state and another follows a highly irregular course; this is because the line follows some mutually agreed divide based upon a particular criterion such as a linguistic boundary, e.g. the boundary between Austria and Yugoslavia and between Hungary and Romania.

30. The boundaries in South America. It will be useful to comment briefly upon the boundaries in a particular area and we will use South America as a case study. In a continent some use is made of natural features for boundaries and, as noted above, both mountain ranges and rivers perform this function. On the other hand, one of the interesting features is the way in which, in most of the Andean countries, the states straddle the Andean system. One might have expected the formidable cordilleran ranges to have provided perfect natural frontiers but, with the exception of Chile, this is never the case.

Another interesting feature is that the political framework has remained comparatively undisturbed. Inter-state conflicts have been fewer than the disturbed political conditions might lead one to expect, and the number of territorial changes which have occurred during the past century and a quarter have been confined principally to one or two areas. The chief frontier disputes and territorial changes have involved Ecuador, Bolivia and Paraguay.

Disputes with her neighbours caused Bolivia to lose her Pacific outlet, thereby making her a landlocked state (*see* Fig. 24). The eastern section of the country, the *Oriente*, was largely

[*After Borchard and Schurz*]

FIG. 24 *Bolivia: boundaries and lost territories.*

The War of the Pacific (1879–83) resulted in Bolivia losing its Pacific coastline. The dispute between Chile and Peru over the Tacna–Arica area was not settled until 1929 when the area was divided: by this same agreement, the port of Arica was to become a free port for Bolivia. Bolivia still hankers after a Pacific outlet and, to solve this unrest, it has been suggested that she should be given a corridor to the coast.

unoccupied, unpeopled and undeveloped: this led to her more rapacious neighbours slowly taking over her eastern territories. Bolivia has learned her lesson: she realises that if she is to keep her territories inviolate she must occupy, develop and populate them; a vacuum is a temptation to all.

CAPITAL CITIES

31. The importance of the state capital. Just as there can be no state without lines of communication to link and unify its several parts and a frontier zone and boundary line to protect it from external interference and mark the limits of its sovereignty, so also no state can exist without a capital: a capital is the nerve centre of the body politic. The capital is the seat of government where the essential function of administration takes place. By virtue of its political significance, the capital city may also become the chief ecclesiastical, cultural, financial and commercial centre of the country, although this does not inevitably follow. In Sweden for example, the capital is Stockholm but the ecclesiastical centre is Uppsala; and in the United States, although Washington is the capital, New York is the chief financial and commercial centre. Since the capital is the nerve centre of the state, disaster threatens a country if in time of war its capital should be captured. The occupation of Paris by the Germans in 1940 was a great psychological setback for the French.

A state capital may exert an influence considerably wider than that of its avowed function as a national administrative centre: it may become, as in the case of Paris, Rome, Cairo and Jerusalem, symbolic of much more.

Capitals are often the largest cities in political units, e.g. Paris, Lisbon, Athens, Cairo, Buenos Aires, Mexico City. This is by no means inevitable, however, for in Canada Montreal is much bigger than Ottawa, in Turkey Instanbul is larger than Ankara, and in Morocco Casablanca is greater than Rabat.

A further point to note is that capitals are frequently associated with the core areas or heart regions of states: this is true of London, Paris, Athens, Moscow and Ankara.

32. Types of capitals. Many years ago Dr. Vaughan Cornish made a study of capital cities and suggested a simple but useful division into four types (*see* **33–36**).

33. Storehouse capitals. These are the cities which lie roughly in the centre of productive natural regions and which form the nuclear areas of states. Because they were regional centres, the communications tended to become focused upon them also. One can think of a number of fairly obvious examples:

(*a*) Paris, which lay in the centre of the fertile and productive Paris Basin, became the natural focus of that basin, more especially since the natural routeways, provided by the river valleys, all centred upon the city.

(b) London, lying in the centre of the London Basin and at the lowest crossing point of the Thames, though ex-centric to the English Plain, was conveniently placed with respect to lowland England and it, too, became a natural storehouse.

(c) Prague, situated approximately in the centre of the "Bohemian diamond" at a focus of routes, was the traditional capital of the old kingdom of Bohemia which, subsequently, became the nucleus of the new state of Czechoslovakia.

34. Forward defensive capitals. There are many cases of capitals lying in forward defensive positions. They cover the approaches to the country, facing the direction whence came the greatest threat. Such capitals usually hold strong strategic situations, commanding important and often easy routes. These are examples:

(a) Edinburgh, whose site was a fortifiable crag, commanded the narrow coastal route to the south, i.e. to England. For many centuries Scotland's greatest enemy was England and so Edinburgh was well placed to protect the approaches to the most important part of Scotland, i.e. the Central Lowlands.

(b) Peking, a great walled city in the north of China, commanded the corridor between the desert and mountains on the west and the sea on the east. Traditionally, the greatest threat to China has come from the steppes of interior Asia; hence Peking was well placed.

(c) Poland offers one of the best examples of the capital occupying a forward defensive position and fulfilling a defensive role. In response to the threat from its neighbours, Poland moved its capital three times. In the earliest days, when the threat came from the Germans, the capital was Poznań in the west; later when the greatest threat moved to the south (i.e. from the Austrians and Turks) Kraków became the capital, and finally, when the greatest threat came from Russia, the capital moved to Warszawa (see Fig. 25).

35. Religious capitals. In earlier times, when religion counted for more than it does today, religious centres frequently grew into state capitals. Some of these religious centres continue still to function as national capitals.

(a) St. David's in Wales and St. Andrews in Scotland are both

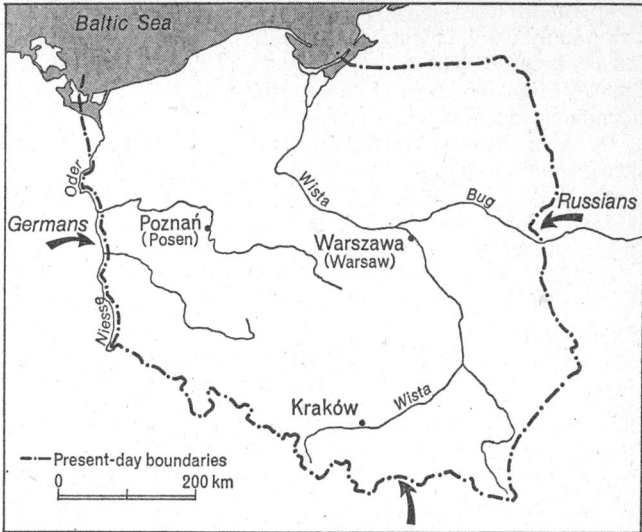

FIG. 25 *Poland's capitals.*

The basin of the River Wista, which lies in the North European Plain, is the traditional homeland of the Poles. Poland has few natural boundaries and has been easy to attack. The Poles have moved their capital three times. On each occasion the move has been to meet the greatest threat of attack. Thus Poland's historic capitals provide good examples of towns having a forward defensive position.

one-time capitals: both began as religious missionary centres. The old capital of Finland was Turku, the first base of Chrisianity in Finland.

(*b*) There are many cases in south-east Asia of religious centres functioning as capital cities, e.g. Kandy in Sri Lanka and Ayuthia (at one time) in Thailand.

(*c*) Tibet, until its recent conquest by China, was a theocratic state and Lhasa, which had the Potala or palace of the Dalai Lama, was the political, as well as the religious, capital of the country.

36. Artificial capitals. These are of two kinds: those created in the past, such as Madrid and St. Petersburg, and the modern administrative capitals, such as Canberra and Brasília.

(a) Madrid was built by Philip II of Spain. He chose the site —naturally an unfavourable one with little to commend it— mainly because of its central location with respect to his dominions. It has since been given an artificial nodality by focusing communications upon it.

(b) After the First World War, Ankara was chosen to be the new capital of Turkey (which no longer possessed its imperial territories) by Kemal Ataturk, the national leader. Ankara replaced Istanbul, the historic capital, because the latter was not representative of the new Turkey.

(c) Canberra was chosen to be the capital of Australia largely because of the contending claims of the state capitals, such as Sydney, Melbourne and Adelaide, to be the national capital. It was decided to solve this rivalry by building an entirely new administrative capital in a federal enclave (see Fig. 26).

37. The new trend of capitals. A very high proportion of the historic capitals have a coastal or river location. Most are by the sea or near to it. This is partly to be explained by the importance of water communications in the past. A few examples, taken at random, illustrate this water siting, e.g. Dublin, Oslo, Lisbon, Athens, Montevideo, Wellington. There are many which are on rivers close to the sea, e.g. London, Paris, Rome, Bangkok and Hanoi. Until recently Karachi and Rio de Janeiro were capital cities.

Particularly since the Second World War there has been a distinct trend towards an interior siting of national capitals. There are several reasons for this:

(a) To provide the state with a more centrally situated capital to assist national unity.

(b) To overcome the excessive, and sometimes corrupt, influence of the old capital.

(c) To distribute more evenly the population and to help open up interior, underdeveloped areas.

(d) To provide a capital which is aesthetically, architecturally and culturally worthy of the state.

NOTE: Some examples of the new capitals which have moved inland are: (i) Islamabad, which has replaced Karachi as capital of Pakistan. (ii) Brasília, which has replaced Rio de Janeiro in Brazil. (iii) Ankara, which has replaced Istanbul as capital of Turkey.

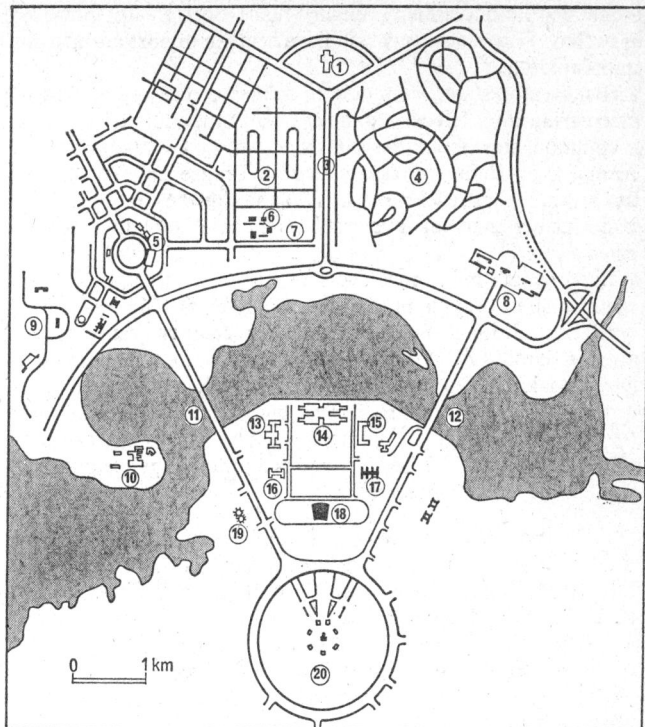

FIG. 26   *Artificial capital: Canberra, Australia.*

Key to numbering: 1. Australian War Memorial; 2. Reid residential area; 3. Anzac Parade; 4. Campbell residential area; 5. Civic centre; 6. Technical college; 7. St. John's Church; 8. Russel offices; 9. University; 10. Hospital; 11. Commonwealth Avenue Bridge; 12. King's Avenue Bridge; 13. National Library site; 14. Parliament House site; 15. High Court site; 16. Treasury building; 17. Administration building; 18. Parliament house; 19. Hotel Canberra; 20. National Centre, Capital Hill.

POLITICAL GROUPINGS

38. Allies and enemies. Rivalries and hostilities between countries have existed since the very beginnings of political organisation: Athens fought Sparta, Rome fought Carthage, England fought Scotland, Germany fought France. Such rivalry, however, as we have suggested, may well have been an important factor in state-building.

Historically, states attempted to achieve a measure of security by entering into defensive alliances with other countries against a common enemy. In the inter-war period Germany was the enemy, a potential aggressor, of both France and Poland and this mutual fear drew France and Poland together. France went a stage further than this and tried to encircle Germany with her allies.

But one of the strange facts of history is that yesteryear's enemies often become present-day allies. Until 1815 France had been the traditional enemy of England; thereafter France became an ally. Until 1945 France and Germany were enemies: today they are friends. During the Second World War the Soviet Union was an ally of the Western Powers but two years after the peace the Soviet Union had ceased to be an ally. All this goes to show that alliances and friendships are temporary expedients largely made to serve nationalistic ends.

39. The free world and the iron curtain countries. Once the common enemy—Germany—had been beaten, both the Allies and the Soviet Union became suspicious of the other party's intentions. There had been disagreements between Stalin, Roosevelt and Churchill over the post-war future of Germany and several of the east European countries even before the war had come to a close; these differences were apparent at the Potsdam Conference after the peace but there was no open disagreement among the victors. Before long an open rift emerged: Churchill referred to an "iron curtain" separating western Europe from Soviet-dominated eastern Europe. Thus Europe came to be divided into two camps: the free democratic countries in the West, supported by the military and economic power of the United States, and the communist countries in the East, dominated by the Soviet Union.

Two defensive alliances gradually emerged:

(*a*) The North Atlantic Treaty Organisation (NATO), set up in 1949, including the United States and Canada and most of the countries of western Europe.

(*b*) The Warsaw Pact, established in 1955, which united the Soviet Union and the other states of the communist bloc, excepting Yugoslavia which had been expelled from the communist community in 1948.

The 1968 invasion of Czechoslovakia by the Soviet Union and her allies resurrected the old distrust of the Soviet Union and tightened the bonds between the NATO countries which, recently, seemed to have slackened; Russian backing of Cuban forces in Angola in 1976 has had a similar effect and in Ethiopia in 1978.

The world conscience over civil rights has been stirred in recent years and the Helsinki Conference and the follow-up conference in Belgrade more recently were attempts to get international agreement on basic human rights. Unfortunately, though participants often pay lip-service to such rights little practical action seems to follow from the Communist bloc countries and currently we are witnessing genocide on a large scale in Africa and south-east Asia, though few countries are prepared to condemn strongly this atrocious inhuman practice.

40. European Economic Groupings. At the end of the Second World War most of the countries of Europe lay in economic ruins. It was generally agreed by the governments of the West that economic help and rehabilitation were essential to prevent communism spreading further westwards (both France and Italy have substantial communist parties). The United States offered immediate assistance under what came to be known as Marshall Aid and in 1948 an organisation was set up to administer this aid: this became the Organisation for Economic Co-operation and Development (O.E.C.D.).

In the 1950s a series of developments occurred which encouraged and aimed at economic co-operation and integration by the countries of western Europe (*see* Fig. 27). The chief stages in this integration were as follows:

(*a*) 1951, the adoption of the "Schuman Plan"; a plan for economic co-operation.

(*b*) 1952, the formation of the European Coal and Steel Community.

(*c*) 1958, the formation of the European Economic Community, frequently referred to as the "Common Market".

(*d*) 1958, the creation of Euratom to develop a common nuclear energy programme.

(*e*) 1959, the formation of the European Free Trade Association (EFTA) mainly instigated by Britain.

The E.E.C. originally comprised the "Six"—France, West Germany, Italy, Belgium, the Netherlands and Luxembourg—

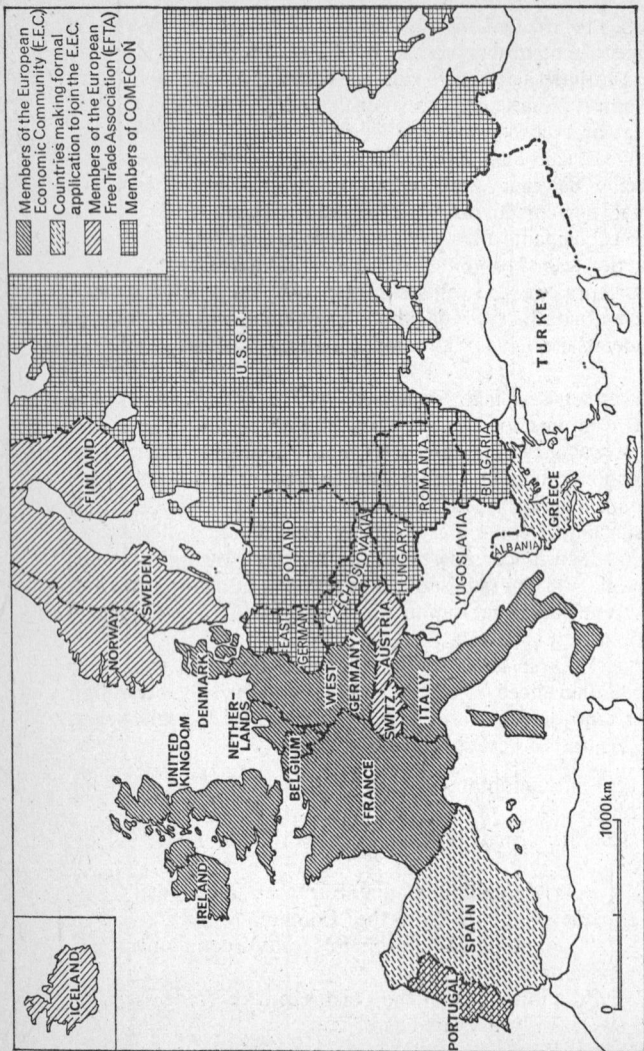

FIG. 27 *European economic groupings.*

Post-war economic chaos and the urge for European unity were largely responsible for the economic associations which emerged in Europe. The strongest of these groupings has been the E.E.C., but even

Legend:

- Members of the European Economic Community (E.E.C.)
- Countries making formal application to join the E.E.C.
- Members of the European Free Trade Association (EFTA)
- Members of COMECON

while EFTA was composed of Britain, Norway, Denmark, Sweden, Iceland, Austria, Switzerland and Portugal, otherwise known as the "Seven". The E.E.C. gave its members access to a large trading area in which the customs barriers were to be reduced in a series of stages. The E.E.C. has proved to be a great success. EFTA, though useful, had not the same possibilities and it has not brought comparable advantages. Britain twice sought to enter the E.E.C. (in 1961 and 1967); on both occasions her application was rejected. Britain joined the E.E.C. in 1973 as did Eire and Denmark.

Many forward-looking politicians hoped that this successful economic co-operation in western Europe would be the prelude to some political association such as a United States of Europe. The institution of the Council of Europe in 1949 held promise of this but largely because of Britain's refusal to agree to any federal authority which might limit her independence of action the Council's function remains very limited.

In eastern Europe the Council for Mutual Economic Assistance (Comecon) was set up by the Soviet Union in 1949. This organisation for economic co-operation and assistance includes the Soviet Union and the east European communist satellite states.

41. Other regional groupings. It is impossible here to review all the regional political and economic groupings but brief reference may be made to a few.

(*a*) *South-east Asia Treaty Organisation* (SEATO) comprising eight countries—United States, Australia, New Zealand, Britain, France, Pakistan, Thailand and the Philippines—was set up in 1954, although Pakistan withdrew from the Organisation in November 1972. Its aim is to co-ordinate economic policies and military planning in south-east Asia. Brought into being by the threat of Chinese communist expansion, the SEATO signatories agree to act in concert should any outsider attack any of their territories in south-east Asia.

(*b*) *The Organisation of African Unity.* Since the Second World War many African states have become independent. With the exception of Nigeria, however, most are poor and weak and underdeveloped. In 1963 the first continent-wide meeting of African heads of state occurred in Addis Ababa. The outcome of this important meeting has led to a Pan-African movement and from it, eventually, may emerge a cohesive bloc of African states,

although it must be admitted that there are many political differences, and rivalries, between some of the individual states. They
share much in common including opposition to the remnants of
colonial rule in Africa, opposition to white supremacy in southern Africa, and economic backwardness. A number of other
regional (largely economic) groupings have emerged since 1963.

(*c*) *The Latin American Free Trade Association* (LAFTA). In
August 1957 the Inter-American Economic Conference, held in
Buenos Aires, took a far-reaching decision: it decided to explore
the possibilities of creating a Latin American regional market.
The establishment of the E.E.C. in Europe was no doubt a
strong factor influencing this decision. Thus in 1960, at Montevideo, seven Latin American countries signed a treaty which
brought LAFTA into being. This development has not only
stimulated inter-regional exchange but has promoted a measure
of mutual self-help amongst the Latin American countries. The
Central American Common Market (1960) and the Andean
Group (1969) are other Latin American economic groupings.

(*d*) *Organisation of Petroleum Exporting Countries* (OPEC).
During the past decade or so the countries of Europe, and
especially the E.E.C. countries, have come to rely heavily on fuel
imports—from 30 to 50 per cent of requirements—and even the
United States now imports 10 per cent of her energy needs. The
oil-producing and exporting countries, such as Saudi Arabia,
Iran, Iraq and the other Middle Eastern countries, together with
Nigeria, Venezuela and Indonesia, banded together to form
OPEC. Because these countries are substantial producers of oil
and have large surpluses for export, they are in a very strong
position to control the international trade in oil. OPEC has put
up the price of oil several times, and, in so doing, has contributed
to the recession in Europe and balance of payment difficulties.
But the energy crisis is not merely an economic one related to the
expense and shortage of oil, for there has been an increasing
tendency for the Arab countries to attempt to use oil as a political
weapon.

THE POLITICAL GEOGRAPHY OF THE OCEANS

42. The law of the sea. The development of political control
over the sea has always followed closely upon man's interest in
the sea and its resources, whether these were rights of passage for
ships, fisheries or, more recently, the mineral wealth on or below

the sea bed. The growing exploitation of the fisheries has raised problems of fishing rights and the conservation of stocks. The recent advances in marine technology together with realisation of the potential mineral wealth that lies on or under the sea bed has focused man's attention upon the oceans and upon the international law which governs national control of the seas.

The first outstanding thinker on the law of marine control was a Dutch jurist, Hugo Grotius (1583–1645), who, in a book, *Mare liberum*, published in 1609, laid down the principle of the freedom of the seas, a principle which has been accepted ever since, until recent years when it has begun to be challenged. Contrary to Grotius' view was that put forward by the English jurist, John Selden, who in his book *Mare clausum*, argued that "the sea by the law of nature or nations is not common to all men but capable of private dominion or property as well as the land". However, Grotius' principle prevailed and came to be generally accepted during the eighteenth century. Although countries accepted the Grotian principle of open seas, they claimed a measure of jurisdiction over the sea adjacent to their shores and this claim was formally recognised in the early years of the eighteenth century when Van Bynkershoek suggested that "territorial waters' should extend as far as guns could shoot from the shore, i.e. about 5 km. This proposal was accepted and until recent times was generally adhered to.

43. The Truman Declaration. As early as 1930, at the Hague Conference for the Codification of International Law, it became apparent that many states felt the old 5 km limit to territorial waters was out-dated, although the legal extension to 19 km did not come until later. Immediately after the Second World War, in 1945, President Truman of the United States announced that "having concern for the urgency of conserving and prudently utilizing its natural resources, the government of the United States regards the natural resources of the subsoil and sea bed of the continental shelf beneath the high seas but contiguous to the coasts of the United States as appertaining to the United States, subject to its jurisdiction and control". This statement was not meant to impair the use of the high seas to free and unimpeded navigation by other countries, but it did imply that the United States looked upon the continental shelf as far as the 200 m isobath was to be looked upon as United States territory. Truman's declaration may be said to mark the beginning of the competition to exploit the sea bed. At the Geneva Convention on the High Seas in 1958 (which came into force in 1964), the Truman Declaration was adopted.

44. Jurisdiction over territorial waters. The 1964 Geneva Convention on the Territorial Sea and the Contiguous Zone provided that the extent of the territorial sea should not exceed 19 km. However, not all countries were signatories to this agreement and where it does not suit them they do not subscribe to it. For example, most of the countries of South America, since they have no continental shelf, claim jurisdiction over territorial waters which extend 320 km from their coasts, while Guinea in west Africa claims up to 210 km, and Gabon in west central Africa up to 40 km. Such extensions are frequently undertaken for fishery purposes. As the need for additional food becomes more acute, the fishery resources become increasingly important and the seas are being fished more assiduously with the aid of new techniques. The North Sea and the adjacent waters of the North Atlantic have long been one of the richest commercial fishing grounds but their over-fishing is leading to declining returns. The herring fishery which once was so important has shrunk to such an extent that there are fears that stocks may not be able to recover from further depletion. The British imposed a ban on herring fishing within a 320 km zone in July 1977 and other E.E.C. fishing countries have followed suit. It seems likely that the fishing for herring will not be able to be resumed perhaps until 1979. Exclusive economic zones, usually up to 320 km are now being claimed by most countries and one recalls the unilateral action of Iceland which led to the "cod war" between that country and Britain.

45. Sub-sea minerals. Man's demand for most minerals continues to grow vigorously and as minerals are wasting assets (exhaustible resources) shortages are beginning to appear. As a result man is beginning to look to the sea bed as a new source of supply as the land-based deposits show signs of exhaustion. Recently there has been much talk of the vast mineral wealth of the sea bed, although so far only a few minerals have been recovered, and apart from offshore mining of coal and oil the others, such as sand and gravel, iron sands, diamonds, and tin, have been recovered by dredging. But the recovery of mineral nodules from the sea bed as a source of manganese, nickel, copper, cobalt, etc. is now being planned. It is these potential riches of the ocean floor that has led to claims for the carving up of the sea bed. The dividing up of some water areas, such as the North Sea, have already been agreed upon.

Two points should be made in relation to these claims upon the sea bed:

(a) The parcelling out of the oceans among those countries possessing a coastline would be a cause for much resentment by those land-locked states left out of any such agreement; they would claim that the riches of the oceans should be a legacy for all mankind.

(b) Deep-sea mining of solid minerals requires advanced technology and one cannot ignore the high cost involved nor the commercial viability of the minerals compared with resources still existing on the land surface. Although one cannot dismiss the enormity of the ocean, mineral resource, progress in its exploitation is bound to be slow.

GEOPOLITICAL CONCEPTS

46. Geopolitics. The term "geopolitics" was invented by a Swede, Rudolf Kjellén, who was the author of some of its main ideas. Geopolitics is based upon an indisputable fact: that the fate of states has been largely dependent upon geographical considerations. During the inter-war period, the Germans, and more especially the Nazi party, accepted and developed some of the geopolitical ideas. A German called Haushofer, who was Professor of Geography at Munich University and subsequently a major-general in the German army, became the leading exponent of geopolitics. He brought the knowledge of the world which he had gained as a trained geographer to help the Nazi regime in their aims at world hegemony. Haushofer elaborated the ideas of German *Geopolitik*, a curious mixture of geography and politics which was pseudo-scientific. Geographers rejected this German *Geopolitik* and its ideas and claims have been discredited.

Nowadays, however, the term geopolitics, as a shortened form of political geography, has gained a new respectability and some people use it as a synonym for political geography.

47. German "Geopolitik". The German geopoliticians of the inter-war period had a twofold part to play in the plans of the Nazis:

(a) They were able to bring to the aid of the German armed forces their specialised knowledge of world geography.

(b) Their theories and publications were used to educate the German people to appreciate the need for, and to support a bid for, world power.

The geopoliticians saw, too, that the next war would be a "total" war, a war quite different in kind from the First World War and all previous wars. Geography, argued the geopoliticians, must be used not merely to help the actual military campaigns to choose the most suitable places for attack and defence, but to study the location of factories, communications, resources and economic strength of enemy countries and to assess the ethnic and geographical ties which might influence military effort and military strength or determine the formation of alliance based on common interests.

48. Haushofer's ideas. It was the special task of the German geo-politicians to provide this economic and military intelligence service and this, it must be admitted, they did remarkably well. But they did more than this. Haushofer produced several theories which were closely connected with Nazi aims and outlook.

(a) Kjellén had put forward the idea that a state was like a living organism which had to grow or perish and Haushofer, accepting Kjellén's idea, said that Germany must expand or perish: without *Lebensraum*—"living space"—the German nation would suffocate.

(b) Haushofer also discussed "pan regions", a cloak for claims upon regions with which Germany had no rational or historical connection. He advocated a German "race area", a "culture area", and a "trade area".

(c) The idea of autarky, whereby the state should attempt to make its economic life as nearly self-contained and self-sufficient as possible, was a further idea that Haushofer and his associates gave to the Nazi politicians.

49. Mackinder's Heartland Theory. Another theory which attracted Haushofer was that elaborated by an Englishman, Sir Halford Mackinder, a one-time Professor of Geography at London University. Mackinder, who wrote a famous book en-titled *Democratic Ideals and Reality*, elaborated upon the idea that geography influenced politics. He showed how geography might explain the general development of world politics, quite apart from the details of military campaigns, particular wars and international incidents. Taking a global view, Mackinder drew attention to the fact that the Old World land mass of Eurasia–Africa was the largest, richest and most populous area on the earth's surface and that it was, therefore, the centre of gravity for human life. The core area of this Old World land mass he termed the "Heartland" and this he defined as "the northern part and interior of Euro-Asia. It extends from the Arctic coast to the central deserts, and westward to the broad isthmus between the Baltic and Black Seas." (*See* Fig. 28.)

Mackinder sought to show that any country which controlled this pivotal area or heartland was in a position to become the

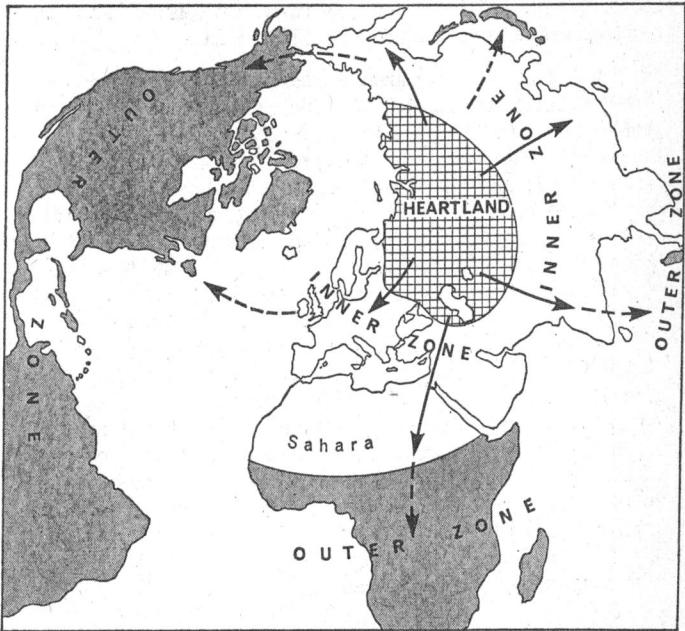

FIG. 28 *Mackinder's heartland.*

The heartland was the "pivotal area' of the Old World land mass. Inaccessible to naval power, it was envisaged as the strategic core area from which a powerful military force could dominate the "inner zone" and thereafter the "outer zone".

most powerful political unit ever and to dominate the world. Any land power which ruled this pivotal area would be easily able to gain control of what he described as the "inner zone" and so to dominate the then two sea-empires of Britain and Japan. When control has been gained of the "inner zone", it would be easy to dominate the "outer zone". To emphasise the strategic value of the heartland, Mackinder wrote (1919):

"Who rules East Europe commands the Heartland.
Who rules the Heartland commands the World Island.
Who rules the World Island commands the World."

Haushofer saw the validity of Mackinder's argument and the Nazi belief in the heartland idea helps to explain the German attack on Russia, the German army's drive towards the Caucasus, and the attempt to gain control of the Near East.

50. The validity of the heartland concept today. At the present time the U.S.S.R. occupies Mackinder's heartland region and since the end of the Second World War we have seen Soviet influence extend westwards into Europe, southwards into the Mediterranean and Near East, and eastwards into Asia. There is no doubt that the Soviet Union occupies a strong strategic position and, as the last war showed, her great area allowed her to develop defence in depth and to disperse her war industries.

Mackinder, however, put forward his ideas before the great developments in aviation and the use of fast, long-range aircraft; also before the development of nuclear weapons and missiles. These developments have altered radically the importance of this heartland position. It can be seen that the heartland is surrounded by encircling peripheral regions and that these regions, if in mutual defensive alliance against a Soviet threat, are in a position to saturate the U.S.S.R. and the heartland region with nuclear missiles. Thus recent techno-scientific developments have completely changed the situation and a geographical location which fifty years ago was a potential source of great political power is now potentially a source of weakness.

PROGRESS TEST 9

1. What conditions are required for the formation of states? (3, 4)

2. Distinguish between the terms state, nation and nation-state. (3, 5, 7)

3. What criteria are used to define nationality? **(6)**

4. Describe fully the meaning of geographical position. **(10)**

5. What advantages has the large state over the small state? **(12)**

6. What is the importance of internal communications to the state? **(17)**

7. Carefully distinguish between the terms frontier and boundary. **(20–23)**

8. Give examples of rivers, marshes, and mountains serving as state boundaries. **(25–28)**

9. What are artificial boundaries? Give examples of the different types of artificial boundaries. **(29)**

10. Suggest a classification of capital cities, quoting examples of each type of capital. **(32–36)**

11. Give reasons for the modern trend towards the interior siting of national capitals. **(37)**

12. Explain the meaning of the following abbreviations: NATO, O.E.C.D., SEATO, EFTA. **(39–41)**

13. What is meant by OPEC? Name is chief members. **(41)**

14. Why are countries claiming territorial extensions over the waters adjacent to their coasts? **(44)**

15. Loftas has described the oceans as the "world's last resource". What does he mean by this and what possible resources have the ocean waters and beds to offer? **(45)**

16. Summarise the ideas of German *Geopolitik*. **(46–48)**

17. Explain what is meant by the "heartland" theory. Is it valid today? **(49, 50)**

CHAPTER X

Recreation and Tourism

THE GROWTH OF LEISURE AND LEISURE ACTIVITIES

1. Reasons for the growth of leisure. During the past 100 years, but more especially since around 1950, there has been an astonishing growth of leisure activity—a development aptly termed by Dower "the fourth wave". Four important reasons underlie the growth of leisure and of leisure activities:

(*a*) Increases in income and higher standards of living, including shorter working hours and longer holidays with pay, have resulted in people in the developed industrial countries having more time and money to spend on leisure. The development of leisure activities is part of the process of economic development.

(*b*) As incomes increase, the amount of surplus income at the disposal of the individual—i.e. that not spent on the essentials of life such as rent, rates, food, clothes, heating—increases, the pattern of the consumer's expenditure changes, and more money is spent on leisure activities, including holidays.

(*c*) Improvements in transport and in communications technology, such as the growth in private car ownership in particular but also in air transport and the decreased cost of air travel (especially through "package tours") have created a boom in both leisure activities generally and in holidaying.

(*d*) There has also developed a new philosophy of leisure activity and holidaymaking (partly stemming from changing social conditions, better education, and the influence of the mass media) which has stimulated active participation in leisure pursuits and promoted a desire to engage in sightseeing and adventurous activities.

There are many factors at work, but these four have been fundamental and largely explain the "explosion" in recreation and tourism.

2. Recreation. To define leisure is not easy: "we are at leisure when we have time free from the necessity to work" (Cosgrove and Jackson, *The Geography of Recreation and Leisure*) is one modern definition. Work here means the daily activity for which we are paid remuneration, for it will be clear that many leisure activities entail work, e.g. gardening.

Recreational activities fall into two groups: (*a*) passive, (*b*) active. Passive recreation includes such activities as reading, watching T.V., strolling round the garden, playing cards, dining and drinking out. All are relaxing activities. Active recreation almost always involves some physical effort—walking, gardening or playing golf, football, or tennis. Active recreation usually occurs out-of-doors and yet passive recreation is the most important form of outdoor recreation. Many watch sporting events, but the greatest demand for passive recreation comes from the car-borne family on a whole-day or half-day outing. There is a strong demand for quiet enjoyment of the countryside or coast, as is indicated by the results of the North West Sports Council's 1970 survey.

However, since 1950 there has been a great expansion in active leisure-time activities. In addition to the traditional participation in games (football, cricket, golf, tennis), there has been a vigorous growth in such activities as climbing, potholing, yachting, water-skiing, horse-riding, gliding, etc.

In times when the average length of the working week is steadily falling, holidays with pay are increasing in length, and the age of retirement is getting earlier, people in Britain generally have more leisure time on their hands than ever before. In view of this and the fact that recreational activities and patterns reflect, to a very considerable extent, the supply of facilities, it is becoming increasingly important that authorities should start to plan and to create the facilities, in both urban and rural areas, to satisfy what certainly will be an increased demand for recreational activities.

3. The tourism phenomenon. Tourism is merely an aspect of leisure activity, though probably its most important aspect. Travel, from very early times, has had a fascination for man: the urge to discover the unknown, to explore new and strange places, to seek new environments and experiences. Travel to achieve these ends is not new, but tourism, as we understand the term today, is of relatively modern origin. Tourism is distinguishable by its mass character from the travel undertaken in the past. The mass movement of people annually from their home location to some other temporary location for a few days or weeks is a product almost entirely of the period following the Second World War. The annual migration of large numbers of people began rather more than a hundred years ago but the present-day exodus, especially in relation to international tourism, is essentially a post-1945 phenomenon.

Over the past thirty years—since the world began to settle down after the years of readjustment immediately following 1945—there has been an astonishing growth in both domestic and international tourism. The United Nations reported that in the ten-year period between 1955–65 the number of tourist arrivals (in some 67 countries) trebled from around 51 million to over 157 million. By 1975 it was around 200 million. Although since 1973, as a result of world economic difficulties, there has been a slackening off in the rate of expansion, prior to that date international tourism had been growing at a rate of about 12 per cent per annum.

4. The growth of tourism. The annual holiday is an important feature of social life in the West. How and why did the holiday arise? The term "holiday" derives from holy days, days associated with religious observances, though nowadays it is used in a secular sense meaning a respite from the routine of workaday life and a time of leisure, amusement and recreation.

Holidays in one form or another are common to all civilisations, whether ancient or modern, though their character has differed greatly. Public holidays were a feature of ancient Rome and the *Saturnalia*—the feast of Saturn—was an occasion when all classes indulged in feasting and frolic, and even the slaves were permitted privileges. In Christian Europe certain days commemorating religious festivals and saints' days became holy days when there was no work, and fasting and prayer. Subsequently, though public and semi-official offices in England frequently closed on certain saints' days, there were no general public holidays until the Industrial Revolution which wrought drastic social, as well as economic, changes. One of these was the introduction in 1871 of the so-called bank holidays. Gradually the idea of a week's holiday emerged. After 1945 many people secured a second week's holiday; now a fourth week's holiday is quite common.

The practice of "going away" for a holiday is of comparatively recent origin, but it can be traced back to the seventeenth century when the medical profession began to recommend the properties of mineralised waters and started to send their patients to the spas or watering places such as Bath and Tunbridge Wells. The discovery of a mineral spring meant potential prosperity and places were quick to capitalise such good fortune. Dozens of spas grew up and the habit of "taking the waters" became very fashionable.

This formed the first stage in the development of holiday resorts. The second stage centred on the movement to the coast which, in turn, led to the growth of the seaside resort. Doctors began to recommend sea bathing for those afflicted by gout and other ailments. Dr. Richard Russell, a Brighton physician, argued that sea water was as effective for medical use as spa water. And so gradually the habit of sea bathing began, at first for its therapeutic value, but soon for pleasure. Royalty began to visit the seaside resorts and this greatly encouraged their development.

At this time, the eighteenth century, few outside wealthy circles ever travelled more than a few miles from their immediate neighbourhood and holidays away from home were the privilege of the well-to-do. Movement about the country was severely limited partly because of expense, but partly because of the difficulties of road transport. The coming of the railways had a profound effect upon the fortunes of the seaside resort: for the first time people could move relatively freely, quickly and cheaply from one place to another. A spectacular expansion in holidaymaking therefore occurred in late Victorian times.

The habit of "going abroad" for a holiday is very much a post-war feature. Prior to the Second World War, few, mainly the well-to-do, went abroad; today millions each year leave Britain for the Continent or travel even farther afield. The Continental holiday had its beginnings in the "Grand Tour" which the sons of gentlemen undertook in the eighteenth century, but its recent growth is the result of a combination of factors: more leisure time, greater affluence of the working population, improved travel facilities, a desire to get some sunshine, and a wish to travel to see foreign places.

5. The nature of present-day tourism. A number of changes are taking place in tourism which are fundamentally changing its nature.

(a) The whole concept of pleasure travel has changed since pre-war days. Foreign travel prior to 1939 was for the affluent, leisured and well-educated who enjoyed travel for its own sake and who were content to enjoy scenery, works of art, and the flavour of foreign places. This concept, however, has been replaced by "tourism"—something altogether different. The present-day traveller has a different kind of background, and his ideas about travel are very different. He comes from a wider social background and his tastes and desires are much more varied; his leisure time is much more restricted and, accordingly, he wishes to pack into it as much as possible.

(b) There has developed what has been aptly termed the "democratisation" of leisure pursuits. For example, winter sports which not so long ago was an activity almost exclusively confined to the wealthy are now enjoyed by many. The "commercialisation" of many hobbies or leisure-time activities such as riding, boating, water-skiing, hitherto rather exclusive pursuits, has made them available to the ordinary man who is interested. Large numbers of people are now also going abroad to participate in the more exciting and exotic activities of mountaineering, underwater swimming, trail-riding, etc.

(c) There has also been the development of what is generally termed "social tourism". This kind of tourism, epitomised in the British holiday camp, not only bypasses the usual facilities provided by the traditional tourist resorts but is responsible for the opening up and development of new areas. Organisations such as the Club Mediterranée cater for large groups of people and offer specially designed low-price accommodation, catering facilities and entertainment. There has also been a big growth in camping and caravanning and many camp and caravan sites provide varied amenities.

THE CHARACTER AND ORGANISATION OF TOURISM

6. The components of tourism. There are certain essential components that make tourism possible:

(a) *Transport.* This is a very necessary condition since tourism involves going somewhere and makes use of trains, coaches, ships, aircraft and the private car in particular. New increased speeds are reducing travelling time, which is important to the tourist.

(b) *Accommodation.* All, in the process of travel or at their destination, require accommodation which provides food, drink and sleeping facilities. The nature of accommodation is very variable, e.g. hotels, motels, inns, boarding houses, hostels, camps, etc.

(c) *Amenities*. The provision of facilities for bathing, boating, sporting activities, recreation, dancing and amusement is an important item in any resort or holiday centre. The demands of the holidaymaker for a wide range of amenities has led to what has come to be known as "development", a matter which has preoccupied resort managements.

Although there are many other factors predisposing towards the development of tourist areas or centres, e.g. scenic attractions, good weather, cultural features, the three elements of transport, accommodation and amenities are fundamental.

7. The organisation of tourism. Strictly speaking, tourism, like recreation, is not an industry: it is an activity; but, in economic terms, it creates a demand or provides a market for a number of quite separate and varied industries. In some areas tourism represents the major part of the market, in others a complementary, but frequently highly profitable, demand for accommodation, catering, transport, entertainment and other services designed largely, perhaps even primarily, for a residential or industrial community.

If we consider tourism in economic terms, i.e. demand (or production) and supply, we can divide tourism into two sectors, the dynamic and the static. Within the dynamic sector fall the economic activities of (a) the formation of the commodity, (b) the motivation of demand, (c) the provision of transport. Translated into practical terms, the dynamic aspects embrace the activities of travel agents, tour operators, transport undertakers and ancillary agencies. The static sector looks after the "sojourn" part of tourism, the demand for accommodation, food, and refreshment in the main, the chief provider of which is the hotel and catering industry, although there are also other ancillary services involved (G. Janata, paper on *Tourism*, Ealing College of Technology).

The organisation and administration of tourism varies widely. In some countries it is closely regulated by the state; in others, it is more loosely controlled and the private sector is important. Nearly all countries, however, have some sort of National Tourist Office (N.T.O.). The degree of governmental control is likely to be greatest in those countries, especially the developing countries, which are just embarking upon tourism development.

8. The character of the tourist industry. For convenience we may talk about a tourist industry although, as noted above, tourism is an activity associated with a variety of industries. Tourism may be described as a "multi-dimensional phenomenon", for many activities each make their own individual contribution to a comprehensive service to tourists.

The industry is primarily a service industry and a large proportion of those actively engaged in it find employment in tertiary occupations, e.g. catering, transport, entertainment. The industry is also marked by a fairly distinct seasonal rhythm; there are few places enjoying an all-year-round trade. The seasonal character implies that casual work and seasonal employment are usually distinguishing features of the industry. In season, however, the industry is labour-intensive. Out of season, much of the tourism plant lies idle and this, of course, is uneconomic. Hence the attempts which are made through the staggering of holidays, out-of-season holidays at reduced rates, special celebrations, conference organising, etc. to extend the season. Anything which will help to lengthen the tourist season will help the industry generally.

SOCIO-ECONOMIC ASPECTS

9. The economic importance of tourism. It is seldom realised that tourism has become the largest single item in the world's foreign trade—currently it is of the order of £20,000m a year. This is big business and clearly the economic value of tourism is very substantial. Tourism can help a country's economy in a variety of ways.

(a) In countries where the industry is well developed, tourism provides employment on a large scale; in some areas or towns it is the main employer of labour. In many rural and marginal regions tourism is valuable since it offers almost the only alternative employment to such primary activities as farming and fishing. Moreover, it is especially important because it is a labour-intensive industry.

(b) International tourism can also assist the balance of payments; for example, in some countries such as Spain, Mexico and Jordan where tourist receipts represent a substantial proportion of the exports—44, 39 and 28 per cent respectively in 1970—tourism is vital.

(c) Equally significant, although less immediately obvious, are the multiplier effects. For instance, it has been estimated that £100m of expenditure by tourists in the United Kingdom creates £300m of total expenditure throughout the economy. Expressed in another way expenditure on tourism supports activity in other industries, e.g. transport, construction, agriculture, furniture, etc. The importance of "secondary" expenditure, which is inherent in tourism, cannot be overestimated.

(*d*) Tourism also aids a national economy in that it can help to develop and revitalise regional economies more quickly than many other industries and, by its very nature, tends to favour peripheral regions which are the very ones needing an injection of economic capital.

10. The social effects of tourism. Along with economic benefits there are social benefits. On the other hand, tourism development may bring social disadvantages. The social benefits stem from the money brought into areas, especially underdeveloped areas, by the industry. The chief advantages are:

(*a*) Employment which is probably the biggest single social advantage that the development of tourism can bring to a region.

(*b*) The provision of infrastructure—communications, power supplies, piped water, shops, hospitals, schools, etc.—all become necessary when an area is developed for tourism; hence the local population also benefit from these developments.

(*c*) The provision of "service industries", e.g. transport, laundries, etc. is also an advantage as the local communities would most probably not be able to support such facilities were it not for the demands of the tourist industry.

(*d*) The provision of recreational facilities, ostensibly for the benefit of the tourist, also benefit the health and welfare of the people living within the region.

(*e*) Tourism may bring cultural contacts which are beneficial. The attitudes and horizons of the indigenous peoples may be changed and widened as a result of rubbing shoulders with "foreigners".

Against the advantages must be set certain disadvantages; among these are:

(*a*) There may well be conflict with local interests. For instance, where the land is devoted to agriculture, tourism development may make unacceptable demands upon the countryside, in terms of space and public access.

(*b*) The development of tourism may adversely affect other economic activities; for example, the relatively high wages paid in the tourist sectors, especially in the developing countries, may lead to many workers leaving agriculture.

(*c*) There may be social problems—the greater the influx of tourists, the greater the risk of inconvenience and loss of amenity and facilities for those resident in the tourist area.

(*d*) Cultural contacts may not always be advantageous. Undesirable attitudes and modes of behaviour may be introduced which undermine and upset the traditional local life and customs.

Some, if not all, of the many problems that arise could be met and perhaps overcome by careful planning.

BRITISH AND WORLD TOURISM

11. The tourist industry in Britain. Although London has long been a Mecca for foreign visitors, it is only within comparatively recent years that tourism in Britain has become a major industry. Table IV shows that a record number of visitors—over 10 million —came to Britain in 1976 and spent £1,627m, an increase of 14 per cent on 1975. Visitors from North America (United States and Canada) totalled 1.9 million (they came either on holiday or business). There were large increases in the numbers of visitors from France and West Germany—in each case over 1 million. The British Tourist Authority estimated that in 1977, Jubilee Year, some 12 million overseas visitors would visit Britain, earning the country some £3,000m in foreign exchange.

As an invisible export, tourism is Britain's biggest dollar earner and has become a very important factor in her balance of payments. Excluding fares, the difference between the amounts which 7.25 million British people spent overseas and which overseas visitors spent within the U.K. in 1976 was a £620m surplus in favour of Britain's balance of payments. In 1977 this surplus was expected to be in the region of £1,200m.

The British Isles have much to offer the tourist: a variety of landscape in miniature as beautiful and as interesting as one can get almost anywhere in the world; a wealth of ancient monuments, e.g. Stonehenge, the Roman Wall; abbeys, monasteries, cathedrals, castles and great country houses in infinite variety: many towns of great historical interest and attractiveness such as York, Chichester, Chester, and Oxford; while London itself, with

TABLE IV.   U.K.'S TOURIST TRADE IN 1976

Country of Residence	Visits		Expenditure	
	Thousands	% Share	£ million	% Share
France	1,171	11.6	89.0	5.5
Federal Republic of Germany	1,104	10.9	110.9	6.8
Irish Republic	721	7.1	54.4	3.3
Italy	281	2.8	34.7	2.1
Belgium/Luxembourg	683	6.8	52.2	3.2
Netherlands	832	8.2	90.7	5.6
Denmark	193	1.9	23.4	1.4
Total W. Europe— E.E.C.	4,985	49.4	455.3	28.0
Norway, Sweden, Finland	574	5.7	90.4	5.6
Spain	254	2.5	47.4	2.9
Switzerland	249	2.5	37.7	2.3
Other W. Europe	308	3.1	66.4	4.1
Total W. Europe— non-E.E.C.	1,385	13.7	242.0	14.9
Eastern Europe	58	0.6	6.7	0.4
Total Europe	6,428	63.7	704.0	43.3
United States	1,490	14.8	267.7	16.4
Canada	477	4.7	91.5	5.6
Total N. America	1,967	19.5	359.2	22.1
Japan	119	1.9	19.9	1.2
Australia & New Zealand	412	4.1	114.5	7.0
South Africa	130	1.3	28.9	1.8
Latin America	156	1.5	39.2	2.4
Middle East	365	3.6	192.8	11.8
Rest of the World	511	5.1	151.2	9.3
Total rest of the World	1,693	16.8	546.5	33.6
All Countries	10,089	100.0	1,627.5*	100.0

* Includes estimate for expenditure in the Channel Islands.
NOTE: Totals may not sum due to rounding.
Source: British Tourist Authority

its mixture of ancient and modern, its theatres and galleries, is one of the most interesting and captivating of capitals.

From the point of view of the British economy, it is important that tourism be encouraged. The British Tourist Authority, set up by the Development of Tourism Act 1969, must be given much of the credit for putting Britain on the tourist map; it is trying hard to stimulate further the tourism business. This will not be easy to achieve in the current economic climate and in view of the growing competition in the tourist industry. The great weaknesses in the British tourist industry lie on the servicing side: more good provincial hotels are needed, many more motels are required, catering needs improving, and more personal service is called for.

12. The tourist countries. Tourism is highly developed in western Europe and in North America, but elsewhere, with a few exceptions, although the potential is there, development is slight. In view of the economic advantages to be gained from tourism, many of the developing countries have tended to seize upon tourism as offering a quick and relatively easy way of promoting economic development and of solving their balance of payments difficulties. But, even if such countries as Algeria, Iran and the Philippines possess the natural resources for tourism development, these in themselves are not enough, for they lack the infrastructure which is essential for successful tourism development. Many of them will, without doubt, be able to build up a thriving tourist industry, but it must be a rather slow process since there are many difficulties to be overcome. Certainly a tourist industry of any appreciable size cannot evolve overnight.

The United States and Canada and the countries of western Europe are the largest tourist generating countries. To these we must probably soon add Japan. Europe, however, is the principal tourist destination in the world, receiving some 60 per cent of all tourists. Table V shows that Spain alone is attracting some 30 million tourists a year! The Americans and the West Germans are the world's greatest tourists: they are also the greatest spenders. To these countries we must now add Japan. Since 1964, when tourist restrictions were relaxed, there has been an accelerating flow of Japanese travelling abroad. In 1973 slightly more than $2\frac{1}{4}$ million travelled abroad, for in that year restrictions on foreign travel were lifted altogether.

Countries such as France, Switzerland and Italy, which already

TABLE V.   NUMBERS OF VISITORS AND GROSS RECEIPTS ($m)
FROM TOURISM IN 1974

	No. of visitors (m)	Receipts ($m)
Andorra	3.00	—
Austria	10.89	2,289
Belgium and Luxembourg	7.48	695
Bulgaria	3.82	198
Cyprus	0.15	38
Czechoslovakia	11.78	—
Denmark	13.84	641
Eire	1.62	254
Finland	4.86	294
France	16.57	2,640
Germany, East	15.23	—
Germany, West	6.95	2,343
Gibraltar	0.14	6
Greece	1.96	436
Hungary	4.66	244
Iceland	0.68	13
Italy	10.19	2,370
Liechtenstein	0.74	4
Malta	0.12	24
Netherlands	2.68	1,033
Norway	1.05	269
Poland	7.90	146
Portugal	2.62	443
Romania	3.83	60
Spain	30.34	3,188
Sweden	0.71	275
Switzerland	6.22	1,500
U.K.	7.94	1,957
U.S.S.R.	3.45	—
Yugoslavia	5.46	699

possess a highly developed tourist industry have the principal problem of maintaining and expanding it. Countries such as Yugoslavia, Tunisia and Morocco, where the industry is already of some considerable significance, need careful planning and appropriate investment if tourism is to become a very important factor in the national economy. Regions such as South America, Africa and south-east Asia have latent possibilities for tourism development, but tourism is likely to be of a rather limited nature. Countries such as Mongolia, Ethiopia and Haiti have possibilities for tourist development which are very strictly limited because their climates are unpleasantly extreme, or because they are rather remote and isolated, or because they lie away from areas of high income and high population density which supply the tourists, or because their low level of overall development is such that they lack even the basic facilities required by tourism. (See *A Geography of Tourism*, H. Robinson).

PROGRESS TEST 10

1. What are the principal underlying reasons for the recent growth in leisure and leisure time activities? **(1)**

2. Distinguish between "active" and "passive" recreation. **(2)**

3. Trace the historical development of tourism. **(3, 4)**

4. Assess the economic importance of tourism. **(9)**

5. What social advantages and disadvantages does tourism bring? **(10)**

6. Write a brief account of the British tourist industry. **(11)**

CHAPTER XI

Geographical Regions

1. The regional approach. The previous chapters have dealt with particular systematic aspects of human geography which have been related to the world in general. The geographer, however, concerns himself not only with systematic studies but with areas also. More especially he has attempted to distinguish areas which are geographically distinctive and show some degree of unity.

Over fifty years ago, A. J. Herbertson, one of the pioneers of modern geography in Britain, devised a system of natural regions, mainly based upon climate, vegetation and position. This idea of regions, possessing distinctive characteristics, has been elaborated and for many geographers the recognition, description and analysis of regions has come to be the core or focusing point of geographical study.

Each of the major geographical regions dealt with in this chapter possesses fairly distinctive and discernible characteristics of climate, vegetation, animal life, soil and human response (*see* Fig. 29). The chief features of these varied geographical environments which exist on the earth's surface will be summarised in what follows and we shall indicate briefly how geography has influenced the life and occupations of people living in these regions.

THE EQUATORIAL RAIN FOREST LANDS

2. Location and extent. The lands of equatorial rain forest occur between approximately 5° N. and S. of the equator but in places they extend appreciably beyond these latitudes. The chief areas are as follows:

(*a*) The Amazon basin and the coastlands of the Guianas.
(*b*) The Pacific coastlands of Colombia and northern Ecuador.
(*c*) The Guinea coastlands of west Africa and the Zaïre basin.
(*d*) The coastal strip of east Africa between 0° and 10° S.
(*e*) The Malay Peninsula and Indonesia.
(*f*) The coastal fringe of northern Australia.

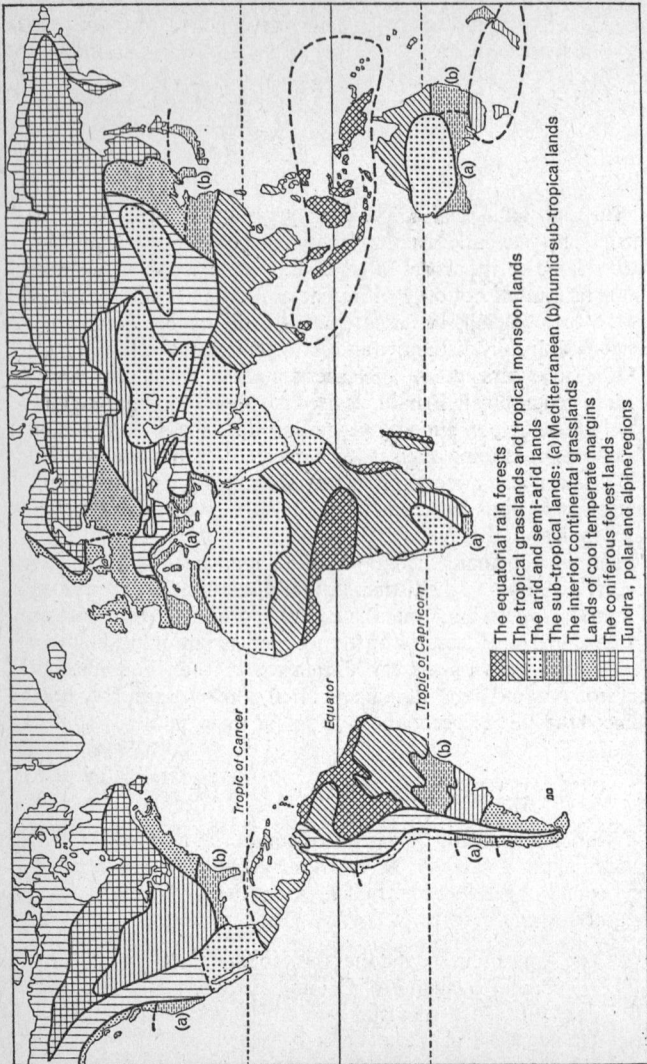

The equatorial rain forests
The tropical grasslands and tropical monsoon lands
The arid and semi-arid lands
The sub-tropical lands: (a) Mediterranean (b) humid sub-tropical lands
The interior continental grasslands
Lands of cool temperate margins
The coniferous forest lands
Tundra, polar and alpine regions

FIG. 29   *Geographical regions of the world.*

Tropic of Cancer

Equator

Tropic of Capricorn

3. Climate. The areas lie in low latitudes and are subject to converging air-streams with, in consequence, low pressures and ascending air and convection rains. As the vertical sun migrates north and south the rainfall does likewise; hence some places have two peaks of heavier rainfall annually. Precipitation, however, is experienced daily throughout the year and is heavy, totalling about 2030 mm. Temperatures are uniformly high, around 27°C throughout the year. The diurnal range of temperature is small but often greater than the annual range, which is seldom more than 2°C. There is no seasonal rhythm but there is a daily rhythm of weather with midday thunderstorms. Except in coastal locations, which benefit from sea breezes, it is generally very humid and oppressive and the climate is enervating.

4. Vegetation and animal life. The perennially hot, wet climate affords ideal conditions for the optimum and uninterrupted growth of plants. The vegetation consists of dense, often tiered, tangled evergreen forest composed of trees of many and widely scattered species. There are many kinds of palms, climbing plants, epiphytes and parasites. Trees grow tall and their well-developed foliage in the upper branches produces a dense canopy; this shuts out light and so ground growth is usually discouraged. The vegetation shows many adaptations to the climatic conditions, e.g. tall, straight, unbranching tree trunks, top growth, buttress roots, large thin leaves with drip-points. Certain shade-loving shrubs, e.g. cacao, grow in the undergrowth. Along low, shelving shores the rain forest, or selva as it is sometimes called, often gives way to tangled mangrove swamp.

Except for a few tree-climbing species, such as monkeys, sloths and jaguars, mammals are few, being largely excluded by plant life, but reptiles, insects and birds abound for they are able easily to move about and find ample food in the nuts and fruits. Creatures are typically arboreal.

5. Environmental handicaps. The equatorial rain forest lands have certain drawbacks from the human point of view and exercise a restricting influence, especially upon peoples who are not very advanced culturally. The more important handicaps are as follows:

(*a*) The unhealthy, enervating climate saps man's energy and favours the development and spread of diseases.

(b) The problems of forest clearance, blights, rusts and plant pests and soil poverty affect agriculture.

(c) The difficulties of pastoral farming due to the absence of indigenous domesticable animals, animal pests and suitable fodder for imported stock.

(d) The difficulties of transport and communication, especially land transport, because of the dense vegetation cover and the problems of maintaining tracks and highways.

6. The influence of environment upon occupations. Primitive, backward peoples, such as the natives of Papua New Guinea or the pygmy peoples of Zaïre, live largely upon the sufferance of nature. They live mostly by collecting the wild products of the forest and by hunting and fishing. In some places a primitive form of agriculture is practised: this is known as tropical shifting agriculture, often called *milpa* cultivation. This consists of clearing patches of forest by ring-barking and burning and then planting seeds or cuttings by means of a digging stick. After a year or two the soil becomes exhausted and a fresh patch of land has to be cleared.

In some areas, where more advanced peoples have colonised the land or established commercial enterprises, or where the natives have come into contact with more civilised peoples, a more advanced type of economy is practised. Plantation agriculture, e.g. rubber in Malaya, hardwood lumbering, e.g. in Amazonia, and mining, e.g. bauxite in Guyana, are carried on. But such activities are fairly localised and there are vast areas of equatorial forest land that remain unexploited.

In most areas it may be said that the forest is *the* dominating feature in so far as man must come to terms with it before he can do anything. The exuberant vitality of plant growth is such that so far man has been largely curbed and confined by it. In most regions the forest forms the chief resource, and upon its control and exploitation will mainly depend the degree to which it forms a home for man and will be of use to him.

7. The influence of environment on life. Civilisation is at a generally low level in most areas although historically in some places, e.g. Java, an advanced culture did emerge; yet even this was more in the nature of an importation than an indigenous development. The native peoples, in general, practise few arts, possess few domestic artifacts, live in simple dwellings which are often little more than shelters, and are organised in small, usually tribal, groups. Seldom do they possess a written language and their beliefs are usually animistic. As a result of the climatic conditions, clothing is scanty, often non-existent. The food supply, contrary to what most people think, is frequently scarce and difficult to obtain and the native peoples sometimes live in a state of near-starvation which is interspersed with periodic gorgings when fortune favours them. The diet is largely vegetal and unbalanced. Diseases are often rife and there is a high mortality rate particularly among children. Notwithstanding the low cultural level of most of the native peoples in the rain forest environment, they often show remarkable ingenuity. In the more open and accessible areas of the forest, and especially where the white man has colonised or intervened, quite drastic changes in the mode of life of the people have occurred; this is probably best illustrated in the instances of Ghana, Malaya and Java.

THE TROPICAL SUMMER RAIN LANDS

8. Location and extent. These environments are situated to poleward of the equatorial rain forest lands. The tropics limit, approximately, their northernmost and southernmost extent. The chief areas are these:

(*a*) The campos of the Brazilian–Guiana Plateau and the llanos of the Orinoco basin.

(*b*) The savanna lands of Africa which form a great horseshoe belt around the Zaïre basin.

(*c*) The tropical grasslands of the northern interior of Australia.

(*d*) The monsoon lands of south-eastern Asia.

9. Climate. The distinguishing climatic feature of these tropical lands is the alternation of wet and dry seasons. Temperatures are always fairly high but especially just before the onset of the rainy season when the sun is at or near the zenith. The sun burns down with great ferocity and the land becomes baked and brown. January temperatures average around 16° C. In the savanna lands the rainfall, which is convectional in type, is confined to the summer months and the total precipitation varies approximately between 500–1525 mm. Although some parts of the monsoon lands may receive light rainfall at other times of the year, the rainfall is principally confined to the summer months (May to September); it is brought by inblowing monsoons which cause heavy, torrential downpours. Rainfall totals are commonly much greater than in the savanna lands. Monsoon regions with less than about 1525 mm are, climatically, very like the savanna lands. In both types of regions the weather generally tends to be sunny, dry, rather windy and dusty in winter and hot, humid and thundery in summer.

10. Vegetation and animal life. In the savanna lands, because of the seasonal rainfall, forest is replaced by grassland. As the amount of rainfall diminishes and the length of the dry season increases, the forest of the equatorial zone diminishes to give a "parkland" scenery, i.e. individual or clumps of trees occur interspersed in the tall grasses. The grasses typically are tall-growing (e.g. elephant grass), coarse, and often grow in tufts. Trees, such as the baobab, acacia and palm, occur widely scattered, except along water-courses.

In the monsoon lands, especially in the wetter parts, there is sufficient moisture to support forest and a deciduous type of forest with many bamboos occurs: but in the drier areas the forest gives way to thorn-bush and savanna.

Animal life falls into two broad groups: the herbivores or grass-eating species and the carnivores or the flesh-eating beasts of prey. On the African grasslands animal life, especially of the grass-eating type, is abundant, e.g. antelope, gazelle, zebra, giraffe, elephant. There are, also, numerous birds. In the other tropical grasslands of the world animal life is much less plentiful. Again, in the monsoon lands, because of the long human occupance of these lands, wild life, other than birds and insects, is not very plentiful, although wild elephants, tigers, leopards, deer and rhinoceroses are to be found.

11. Environmental handicaps. There are certain notable environmental handicaps which are chiefly related to climatic conditions and soils.

(a) The prolonged dry season limits tillage and crop growing is largely restricted to half the year unless irrigation is practised.

(b) In the monsoon lands where the rainfall is variable and chancy, irrigation must be used if crop growing is to be successful.

(c) The long dry season is a severe hindrance to pastoralism for the herbage withers and the water-courses often dry up.

(d) The natural grasses tend to be coarse and lacking in nutritive value and this militates against successful animal husbandry.

(e) Insect pests attack pastoral animals such as cattle, e.g.

the cattle tick, and sometimes, in the case of the tsetse fly, prohibit the keeping of cattle and horses.

(*f*) The soils, subjected to seasonal wetting and drying, become leached and often develop a hard-pan which limits their fertility and usefulness.

12. The influence of environment upon occupations. Amongst backward tribal groups human needs are supplied by collecting, hunting, herding or primitive agriculture, or by a combination of such pursuits:

(*a*) The Australian aborigines live largely by collecting and hunting.

(*b*) The Masai of east Africa are essentially herders of cattle.

(*c*) The Kikuyu of Kenya are hoe cultivators.

In those tropical grassland areas which have been colonised by Europeans, man has devoted himself, so far, principally to cultivation and pastoralism, and especially the latter. Cropping is handicapped by seasonal lack of rainfall but crops which will grow within six months, such as cotton, tobacco and certain oil-yielding plants, are frequently cultivated. Often, however, irrigation is necessary for successful agriculture; for example, El Gezira, in Sudan, is an important cotton-growing area but it is dependent upon artificial water supplies. Soils, too, often are of limited value owing to calcification and they are soon exhausted. Mostly, however, the savanna lands are devoted to cattle rearing, in spite of the drawbacks mentioned above. The industry, except in a few areas (such as northern Australia), is not very flourishing since the cattle are of low grade and the meat poor in quality. The effect of tsetse fly infestation in east and central Africa which severely limits cattle raising should be noted.

In the monsoon lands a sedentary type of agriculture is typical but it varies from a primitive, backward kind to the intensive subsistence type usually termed oriental intensive subsistence cultivation. Except in a few areas of abundant rainfall, great emphasis is laid on irrigation. Rice is the characteristic crop of the wetter areas; where it is drier, millets of various kinds predominate. Jute and cotton are important commercial crops. Some of the drier regions are under savanna and here, if irrigation is absent, the seasonal grasses support a livestock industry as in the Rajasthan region of India and the Khorat Plateau of Thailand. Where there is monsoon forest, the timber industry is often of importance, e.g. the teak of Burma and Thailand.

13. The influence of environment on life. Except where European impact has brought changes, the peoples of the savanna lands continue to be backward and their level of culture is low. In Africa, for instance, society is still largely tribal and this has worked against national cohesion, e.g. in Nigeria. The village is the characteristic settlement type, e.g. the African kraal, but in some areas towns are growing quickly, e.g. in northern Nigeria. Village dwellings are commonly constructed of wicker and thatch. In some parts of Africa, and more particularly where there are towns, dwellings are built of sun-dried adobe or bricks.

The civilisation of the monsoon region is very much a "vegetable civilisation" in the sense that it depends mainly upon vegetable products not only for most of its food supplies but for the raw materials of its manufactures, constructional industries, tools, implements and utensils, household furniture and equipment and frequently for fuel. The food supply is limited in kind, quantity and quality; usually the diet is ill-balanced and lacks the minimum calorie content for physical well-being. Under-nutrition and malnutrition are common and this results in deficiency diseases such as kwashiorkor. Birth rates and death rates are high and life expectancy is short. Abject poverty is the condition of millions. Geographical influences are clearly reflected in food, clothing and shelter. The monsoon lands, however, have a long history of human occupancy behind them and very complex societies and culture patterns have grown up. Language and religion, together with other manifestations of culture such as art, architecture, boat-building, costume and social customs show the rich diversity of culture in the monsoon lands. Unfortunately there is not space to elaborate here but the student who is interested should consult the writer's *Monsoon Asia* (Macdonald & Evans).

THE ARID AND SEMI-ARID LANDS

14. Location and extent. The dry lands occur in tropical and subtropical areas and in continental interiors. It is usual to divide them into two:

(*a*) The hot deserts which lie astride the tropics and which are under the influence of subsiding air masses and out-blowing trade winds.

(*b*) The warm deserts or temperate deserts which lie in continental interiors and are under high pressure conditions in winter and, therefore, out-blowing winds.

Note the existence of a wide belt of arid and semi-arid land which stretches obliquely across the Old World land mass from the Atlantic coast to Mongolia.

The main desert areas are: the Sahara desert; the Arabian–Syrian desert; the Iranian desert; the Thar desert; the Turkestan desert; the Takla Makan desert; the Gobi desert; the Australian desert; the Kalahari desert; the Atacama desert; and the deserts of south-western U.S.A. and north Mexico.

15. Climate. In the hot deserts there is no significant rain at any season. Deserts are sometimes defined as areas having under 254 mm of rain per annum; this is, in general terms, acceptable because as a result of the high evaporation rate the effective rainfall is smaller. Moreover, such rainfall as there is tends to be spasmodic, falling in heavy showers. Some areas, of course, experience extreme drought. The areas are dry because they lie in regions subject to subsiding and outward-flowing tropical continental air.

Very high temperatures are experienced in summer when the sun is overhead—around 33° C. Near Tripoli a temperature of 58° C has been recorded. The mean annual range of temperature is greater than in the other hot regions, while the diurnal range is great owing to the lack of cloud. Nights may be distinctly cold.

The temperate deserts of continental interiors have an extreme climate: summers are hot but winters are cold and, sometimes, very cold. Total annual rainfall is less than 250 mm. In winter, precipitation may take the form of light snow.

16. Vegetation and animal life. Because of the aridity plant growth is sparse. With the exception of annuals which may suddenly spring to life, grow rapidly and die, plant growth is slow. Plants are adapted in various ways:

(a) They may be drought-enduring, e.g. creosote bush.
(b) They may be drought-evading, e.g. annuals.
(c) They may be drought-resistant, e.g. acacias.
(d) They may be succulents which store water, e.g. cacti.

Bush plants have small, tough, leathery leaves which are often reduced to spikes. Root systems are often well developed. Grasses, when they grow, are tough and sporadic.

While the rainfall of temperate deserts may be no more than that of the hot deserts, evaporation rates are much lower; hence the vegetation tends to be more continuous and varied. Bunch grass is common and there are many kinds of shrubs.

Because of the absence of food, animal life is scanty. There are a few species of browsing animals, e.g. gazelle, oryx, some reptiles and many insects. The camel should be mentioned because of its wonderful adaptation to the desert environment. Noteworthy, too, is the locust which afflicts arid and semi-arid regions.

17. Environmental handicaps. Handicaps virtually resolve themselves into a lack of water, for this deficiency prevents tillage (except where irrigation can be practised), restricts natural pasture and affects soils (producing salt accumulation). Though the burning sunshine, great heat and occasional dust-storms may be unpleasant, it is the acute shortage of water which is the major obstacle to human life. Another drawback in some cases is the interior location which renders the arid lands relatively inaccessible. It should be remembered, too, that many desert areas consist of bare rock or pebbles and so have practically no use at all to man.

18. The influence of environment upon occupations. Apart from a few clusters of miners, e.g. the oil-men of south-west Asia, the nitrate workers of northern Chile and the gold-miners of southwestern Australia, the deserts support three main groups of people:

(a) Primitive hunters and collectors, e.g. the aborigines of the Australian desert and the Bushmen of the Kalahari in South Africa.

(b) Nomadic herders, e.g. the Bedouin, Tuareg and other groups of the Saharan–Arabian zone who depend upon the keeping of camels, sheep and goats.

(c) Cultivators, dependent upon oasis wells and surface rivers for their essential water supplies, with which they grow cotton, dates, sugar-cane, cereals and vegetables. Large irrigated valleys include those of the Nile, Tigris–Euphrates, Indus and Colorado. Among the smaller valleys are the "little Egypts" of Peru.

Shortage of water not only affects agriculture: it severely handicaps such activities as mining. Water has to be obtained by sinking deep wells or piping it over long distances, as happens at Kalgoorlie in Western Australia.

19. The influence of environment on life. Except where there is sufficient water to maintain sedentary farming or where mining activities are artificially supported, man must be nomadic. This has meant that he could have no permanent settlement and his few possessions had to be light, portable and unbreakable. His home was usually a tent of skin or felt, his food largely milk, cheese, dates, and occasionally flesh when he killed an animal. His wealth largely resided in his flocks and herds. Because the nomad was a wanderer he was also, often, a carrier of trade; and, also, sometimes a robber and marauder. The stark, bold and unrelenting character of the desert landscape and the strong contrast between sunlight and shade seem to have had some influence upon man's behaviour, ideas and art. The traveller is always welcome and treated with courtesy.

In these lands of great empty spaces and where man had much time to sit and ponder about the mystery of life and the universe, it does not seem surprising that monotheism should have developed here. That astronomy, land survey and mathematics also developed here undoubtedly was helped by the clear, starry night skies and the regular seasonal flooding of the rivers, which also obliterated man-made boundaries. Artistic design tends to be abstract and geometrical and even in the temperate desert, where much of the art is associated with animals, artistic forms are highly stylised.

THE SUB-TROPICAL LANDS

20. Location and extent. These lands lie, as their name implies, poleward of the tropics and extend to approximately 40° N. Strictly speaking, the temperate deserts fall within these latitudes or just to poleward of them but we have dealt with them in the previous section under the arid and semi-arid lands. Here we are concerned with those lands which lie on the western and eastern margins of the continents in sub-tropical latitudes: they fall into two fairly clear-cut types of region:

(*a*) The so-called "Mediterranean" lands:
    (*i*) The coastlands of the Mediterranean Sea.
    (*ii*) Part of the state of California.
    (*iii*) Central Chile.
    (*iv*) The southern tip of South Africa.
    (*v*) South-west Western Australia and south South Australia.

(b) The humid sub-tropical lands:
   (i) Central and southern China.
   (ii) South-eastern United States.
   (iii) The south-east of Brazil.
   (iv) Natal in South Africa.
   (v) New South Wales in Australia.

21. Climate. The climates of the western and eastern margins of the continents in sub-tropical latitudes are markedly contrasted in spite of their common latitudinal position. While temperature conditions are roughly similar, there is a great difference in the amount and occurrence of rainfall and this results in quite different climatic regimes.

The "Mediterranean" lands are lands of marked seasonal changes. Because of their "transitional" position, they come under the influence of tropical continental air masses in summer and of polar maritime in winter: this results in hot, dry summers and mild, wet winters. Average summer temperatures are around 24–27° C; winter temperatures around 10° C. Cyclonic storms bring heavy but infrequent rain in winter, the amount varying between approximately 380–1050 mm. Summers are usually rainless. The weather is sunny with little cloud even in winter. Equable conditions prevail.

The eastern margins are mainly under the influence of the tropical maritime air mass. The climate is characterised by no cold season and no dry season. Monsoonal tendencies are strong and this leads to a summer maximum of rainfall. The total annual rainfall is usually between 750–1550 mm. The weather is variable in winter with showery rains; in summer it is very warm and humid.

22. Vegetation and animal life. Because of the differences in rainfall (total amount and occurrence), the vegetation of the two types of regions differs strongly. Both are regions of natural forest: sub-tropical evergreen in Mediterranean lands and mixed evergreen and deciduous forest in the warm temperate regions.

Mediterranean vegetation is extremely varied, and very interesting in its response to summer drought. In some areas, but particularly in the lands around the Mediterranean Sea, the original forest has been almost completely destroyed and its place has been taken by a secondary growth of bush and shrub vegetation known as *maquis* in France, *macchia* in Italy, *matorral* in Spain, and *chaparral* in California.

The adaptations of plants to meet the dry summers include the following:

(*a*) Many bulbous and tuberous plants which grow in spring and then die back.

(*b*) Small, thick leaves, sometimes spiny or prickly, with wax coatings to check transpiration.

(*c*) Thick barks on trees which often have a compact growth.

(*d*) Wide-spreading roots or deeply penetrating tap-roots to enable plants to find sufficient moisture.

(*e*) Trees often have cones or nuts instead of fleshy fruits; fruit exhibits thick, tough and waxy skins, e.g. pomegranate, lemon, fig, olive.

In the eastern warm temperate forests the vegetation is more luxuriant, trees are taller, and there is an even greater variety of species. In the wetter areas, e.g. Florida, Natal, a rain forest not unlike tropical rain forest flourishes.

Animal life, especially in the long-settled areas of both types of regions, is not abundant. Monkeys, rodents, birds and insects are typical creatures but in Mediterranean lands such domesticated types as asses, sheep, goats and oxen are most common while buffaloes and pigs are dominant in China.

23. The influence of environment on occupations. The traditional crops of the Mediterranean lands are wheat, vine and olive. Tree crops have always been important because of the problems created by summer drought. The dry summers have meant that irrigation was desirable if not always necessary. Today horticulture dominates the economic life of the Mediterranean regions, although field crops are not absent. The olive and vine are widely grown. Citrus fruits—oranges, lemons, grapefruit—require irrigation but are important. Early flowers and vegetables are cultivated in many parts. Few cattle are kept owing to the dearth of nutritious grasses, but goats and sheep are common, more particularly in the lands around the Mediterranean Sea; it is here that transhumance, the seasonal movement of animals in summer up to the mountain pastures, is practised. The attractive climate of these lands has stimulated an important tourist industry, especially in the Mediterranean lands and California.

The economy in the warm temperate eastern margin lands varies widely:

(*a*) In China intensive cultivation by peasant farmers prevails;

emphasis is upon rice cultivation but tea, sugar-cane, vegetable oils, cotton and a wide variety of other crops are grown.

(b) In the United States plantation-type crops, such as cotton and tobacco, are important but maize is grown and in the flat alluvial coastlands some rice. There is, also, a varied industrial development based upon oil and natural gas, timber and cotton.

(c) In south-eastern Brazil a prosperous agriculture based on coffee, sugar, cotton, maize and fruits together with the exploitation of the Parana pine forests and a developing manufacturing industry prevails.

(d) In Natal agriculture prevails and is largely concerned with the cultivation of sugar-cane, wattle and fruits. There is some industrial development around Durban and Newcastle.

(e) New South Wales has a well-developed mixed economy: crop cultivation, cattle farming, and considerable manufacturing activity, based on the Newcastle coalfield, all take place.

24. The influence of environment on life. Both the Mediterranean lands and China were early centres of civilisation; in both areas organised societies developed a high level of culture. The cultivation of field, orchard or garden crops provides a living for a high proportion of the population. Outside these two areas the warm temperate lands are of relatively recent settlement but most of them are also much concerned with farming, although there is growing industry. The seasonal climatic regime has induced man to give forethought and to show care in his work on the land; given this, he gets good returns. Because winter temperatures are seldom low, clothing and shelter are much less important than in cooler climes. Man spends much of his time out of doors whether he lives in Italy or New South Wales. In the "new" lands which have been settled by Europeans the economic and social patterns of western civilisation have been introduced although these have often become slightly modified to suit local conditions. Differences in terrain, climate, resources and opportunities are likely to lead to different adaptations.

THE INTERIOR CONTINENTAL GRASSLANDS

25. Location and extent. These regions are principally found in the interiors of continents in mid latitudes. They occur either in lowland areas or in highlands up to about 1,500 m. The chief areas are these:

(a) The steppes of Europe and Asia.
(b) The prairies of North America.
(c) The pampas of Argentina and Uruguay.
(d) The veld of South Africa.
(e) The temperate grasslands of Australia.

26. Climate. These lands are marked by strong seasonal temperature contrasts; the range of temperature varies according to the distance from the sea and the altitude. The steppes and prairies have greater extremes than the regions in the southern hemisphere. Temperatures are very low in winter in the northern hemisphere: November to February are typically below freezing point. In the southern hemisphere mid-winter temperatures are around 7° C. It is very hot, by day, in summer. The total annual precipitation is low, between about 250–500 mm. About a quarter falls in winter, usually in the form of light snow; in spring and early summer thunderstorm convection rains occur. The weather tends to be calm, clear and frosty during the long winter but strong, keen winds may sometimes blow; summers are warm and bright with occasional rains but they are of relatively short duration. The exception to the general form is the pampas region of Argentina and Uruguay where, at least near the coast, rain is more ample and falls throughout the year.

27. Vegetation and animal life. Generally speaking, the grasses of the temperate grasslands are sweeter, finer and shorter and often more continuous than those of the savannas, but in the drier areas the grassland degenerates into scrub. In summer, the grasses wither and die and the landscape takes on a tawny cast. Trees are few and normally confined to watercourses, but in Australia the landscape is often dotted with scrub from which dwarf eucalypts and acacias rise. The grass grows tallest in areas of heavier rainfall and in the humid pampas of South America tall feathery pampas grass grows.

One can imagine these grasslands, before man took possession of them, being well populated with wild life, much as the North American prairies were with bison before the white man arrived. The animal life consists of grass-eaters such as antelope, wild horse, sheep and goats and kangaroos (in Australia) and guanacos (in South America); beasts of prey such as the puma and wolf; and running birds such as the South American rhea, the South African ostrich, the Australian emu and the prairie hen. Today, of course, these grasslands are the habitat of domesticated animals: sheep, cattle, horses and camels.

28. Environmental handicaps. The principal handicaps of the temperate grasslands from the human point of view are the following:

(a) The shortness of the growing season, particularly in the northern hemisphere.

(b) The light rainfall which is often marginal or insufficient for crop growing.

(c) The dying back of the herbaceous cover in late summer which sometimes poses a problem for the pastoralist.

(d) The distance of the grasslands, more particularly in the northern hemisphere, from the coasts.

29. The influence of environment on occupations. Man is chiefly occupied by the rearing of animals. In earlier times pastoral nomadism dominated the Eurasiatic grasslands but the twentieth century has seen the virtual disappearance of this kind of life. While the Asiatic steppes are still occupied in a few areas by nomadic herding, elsewhere large areas within each temperate grassland region have either been planted with specially selected grasses to improve the natural pasture, and to encourage the commercial production of beef, wool and mutton by settled stock-rearers, or else they have been ploughed up and their fertile soils sown with wheat, barley, maize and other cereals. In many cases the open ranges have been divided into large farm units which specialise in the production of animal and vegetable commodities for western European markets. These are the lands of large-scale extensive cereal cultivation, where agriculture is highly mechanised but output per hectare is relatively low.

It is interesting to note how the temperate grasslands have developed, at least in some areas, through four distinct phases:

(a) Originally, when the grasslands were inhabited by wild animals and man was in a primitive stage of culture, first hunting and then simple herding (following upon the early domestication of animals) took place. People in this phase were essentially nomadic, e.g. the Plains Indians of North America who hunted the bison.

(b) When, later, these grasslands were opened up and settled by more advanced peoples, e.g. Europeans in Argentina in the nineteenth century, pastoralism was continued but undertaken by different methods; large ranches and sheep farms, based on fixed homesteads, replaced the traditional nomadic rearing of animals.

(c) In the more moist and fertile areas the grasslands were ploughed up and sown for cereal crops, chiefly wheat and barley, and other temperate crops. This development usually occurred around 1875 or just after. More recently some of the drier areas have been cultivated by using new farming techniques such as irrigation and dry farming.

(d) In more recent years industry has begun to appear in the more advanced of these regions, e.g. in the Soviet Union, in southern Alberta in Canada. Most of the industries are based upon agricultural products, e.g. flour milling, meat canning, leather working, the manufacture of cereal foods, starch, etc., but where mineral fuels have been found, as in Alberta, a more highly developed industry has emerged.

30. The influence of environment on life. As we have already noted, man in earlier times based his life upon the animals of the grasslands and he was a nomadic herder. Nomadism meant that man could not have a permanent home. He lived in a tent which was often made from the skins of his animals. His household equipment had to be light, unbreakable and portable and was limited to a few items such as bottles, cooking pots, blankets and rugs. His goods were mostly made of leather and wool. His diet was limited and consisted chiefly of milk, cheese and occasionally meat provided by his animals and cereals bartered in the marketing centres he visited periodically. Society was tribal and patriarchal and each group had its own, generally recognised pasturing grounds.

Stone and timber are usually absent in the temperate grassland regions and today where permanent settlements exist, whether towns or villages, most of the building material has to be imported.

LANDS OF COOL TEMPERATE MARGINS

31. Location and extent. These regions are to be found on the western and eastern margins of the continents in temperate latitudes; approximately between 45 and 60° latitude. The chief areas are as follows:

(a) North-western Europe.
(b) Manchuria.
(c) British Columbia, Washington and Oregon.
(d) North-east U.S.A. and south-east Canada.
(e) Southern Chile.
(f) Tasmania and New Zealand.

32. Climate. A distinction must be drawn between the climates on the western and eastern margins; the former have essentially equable conditions, especially near the coasts, while the eastern margins tend to have rather extreme conditions.

On the western sides, the climate is equable with no dry season. Typically winters are mild, summers cool; the warmest month is under 22° C, the coldest above freezing, though frost occurs, especially at night. There is a relatively small annual range of temperature. There is rainfall, mainly of the cyclonic and relief type, all the year round, usually with a winter maximum. The amount of precipitation, which includes occasional sleet and snow, is very variable, approximately between 500–2,000 mm. The weather is changeable, damp and humid with much cloud, although there are fine spells of longer or shorter duration.

In the lands on the eastern margins summers are appreciably warmer but winters are pronouncedly colder. There is a fairly evenly distributed rainfall with a summer maximum; the precipitation averages about 500–625 mm with snow in winter. The weather is characteristically brisk, calm, clear and cold in winter but rather warm and humid in summer.

33. Vegetation and animal life. Deciduous forest is typical but large areas of north-western Europe and north-eastern U.S.A. have been denuded of their forest cover, largely for agriculture. The chief trees—oak, ash, beech, birch and elm in Europe, the oak, hickory and maple in North America, the eucalypts in Tasmania—shed their leaves in autumn and rest during the winter months. At high altitudes conifers often replace deciduous species. The fields of grass, which seem so typical of these lands, are essentially artificial, being man-made.

Formerly rich in animal life, including deer, ox, pig, fox, wolf, bear, the long-settled lands have now lost most of their wild life through progressive extermination.

34. The influence of environment on occupations. Where the forests have been cleared, intensive commercial agriculture usually of the mixed type, is often practised. In areas which are more particularly cool, cloudy and wet, arable farming is handicapped; on the other hand such conditions favour pastoral farming, especially dairying. Dairying and market-gardening usually assume importance near densely populated industrial areas, while farther away from the urban areas arable cultivation is generally prominent. The chief crops are wheat, barley, oats, rye, potatoes and sugar-beet, while in the more sheltered, sunnier areas hard fruits, such as apples and pears, and soft fruits are grown.

By an accident of nature some of these areas are rich in coal and other mineral deposits and, because most of these areas are highly developed, they have become great centres of industrial manufacture:

    (*a*) North-western Europe:
        (*i*) The coalfields of Britain.
        (*ii*) The Franco-Belgian coalfield area.
        (*iii*) The Ruhr of West Germany.
    (*b*) The United States–Canada:
        (*i*) The New England region.
        (*ii*) The Pennsylvania coalfield region.
        (*iii*) The Lake Peninsula of Canada.
    (*c*) Southern Manchuria, around Mukden and Anshan.

In the more recently settled areas, especially of the southern hemisphere, the exploitation of the natural resources, along with farming, characterises the economy:

    (*a*) British Columbia is much concerned with fishing, forestry and mining.

    (*b*) Southern Chile is still little exploited but there is some mixed farming and forestry.

    (*c*) Tasmania is mainly concerned with mixed farming and a little forestry and mining.

    (*d*) New Zealand is principally a dairying and sheep-rearing country with developing industry.

35. The influence of environment on life. These regions, which are neither too hot nor too cold and which enjoy variable weather conditions, have been acclaimed as being climatically the best for human activity, stimulating physical and mental effort. Seldom, too, do the climatic conditions prevent man from working out of doors. If the correlations between maps showing regions of stimulating climate and areas of highly developed civilisation are acceptable, this implies that good health and energy promoted by such a climate must be counted among the conditions necessary for human progress and, therefore, that the high degree of progress and civilisation in most of the cool temperate regions is not surprising.

Because these regions have highly developed agriculture, industry and commerce, the peoples enjoy the highest standards of living in the world; they are the best fed, clothed and housed of any of the world's peoples. The continual changes in the weather, on the other hand, make a well-balanced diet, suitable clothes and substantial dwellings necessary.

Finally, because manufacture and commerce are highly developed these are regions which, generally speaking, have a well-developed urban life. In Europe and the U.S.A. more than 75 per cent of the people live in towns and a high proportion of them in great cities or conurbations of over one million inhabitants.

THE CONIFEROUS FOREST LANDS

36. Location and extent. These occur in sub-arctic regions. In North America the forest extends as far as 40° N. on the west coast, but in the centre and east the 50° parallel marks its approximate southward extension. In Eurasia the southern limit is about 60° N. in the west but around 40° N. in the east; in Central Asia the forest penetrates southwards to about the 50° parallel. There are two main belts:

(*a*) In North America, stretching from British Columbia through the northern parts of the Prairie Provinces, much of Ontario and Quebec to the Maritime Provinces and Newfoundland.

(*b*) In northern Eurasia extending from Scandinavia, through Finland and northern Russia and Siberia to the Bering Strait and the Sea of Okhotsk.

The coniferous forest does not really appear in the southern hemisphere, for none of the continents extend sufficiently southwards.

37. Climate. Summers are short but days are long; in winter the reverse is the case. Temperatures in winter are low, below, and often well below, freezing point. Summers are often quite warm, especially in interior situations, and the weather is pleasant and sunny. The winter is usually cold, crisp and calm with occasional blizzards. The greater part of the scanty precipitation, which is about 200–300 mm, except in western coastal localities where it is greater, falls in the form of snow. Convectional rain showers occur in summer.

38. Vegetation and animal life. The northern or boreal coniferous forests are frequently referred to as the taiga. In contrast to the equatorial forests, the coniferous forest contains relatively few species: the chief trees are the pine, fir, spruce, tamarack, balsam and larch, together with the deciduous birch. Though poor in species, vast stands of the same species occur.

The trees of the coniferous forest generally grow straighter, taller and closer than those of the deciduous forest. The low mean annual temperature necessitates a twelve-month cycle of growth: hence the trees are usually evergreen. The trees show notable adjustments to the climatic conditions: leaves are often reduced to needles to reduce transpiration, cones replace fleshy fruits, the compact shapes of the trees resist strong winds, while the conical shapes help to shed winter snows.

The taiga is the home of many fur-bearing animals, e.g. bear, fox, marten, mink, squirrel, beaver and musk-rat. In Canada caribou are found in large numbers in parts of the forest. In Asia and Europe their place is taken by the reindeer, which is largely domesticated.

39. Environmental handicaps. These coniferous forest lands have certain drawbacks from the human point of view:

(*a*) The long cold winters and brief summers militate against agricultural activities.

(*b*) The podsolic soils developed under coniferous forest are poor, sour and leached.

(*c*) Large areas of the coniferous forest lands are swampy, e.g. the muskeg swamps of Canada.

(*d*) Many areas of the forest, especially those in continental interiors, are remote, isolated and inaccessible.

(*e*) The long, extremely cold winters are not conducive to human settlement.

40. The influence of environment on occupations. Population is mainly engaged in lumbering and trapping. The coniferous forests are the source of the world's chief supplies of softwood timber for pulp and constructional work. The pulp is used in the manufacture of paper, cardboard and rayon. So far, the vast timber resources of these forests have been exploited only in the more accessible areas. Formerly these forests were cut down without any thought of conservation; nowadays re-planting is undertaken in most areas to ensure future supplies. The chief lumbering areas are the following:

(a) British Columbia.
(b) Ontario–Quebec.
(c) Maritime Provinces and Newfoundland.
(d) Scandinavia–Finland.
(e) Northern European Russia.
(f) Manchuria.
(g) Northern Japan (especially Hokkaido).

The coniferous forest lands have numerous fur-bearing animals, hence they have long provided the bulk of the world's valuable pelts. Hunting and trapping are much less important than formerly, partly because furs are less fashionable than they used to be, partly because of the development of commercial fur-farming, and partly because of the use of imitation furs made from rabbit-skins and the like or even from synthetic fibres (nylon fur).

As mentioned above, climate and soils militate against farming and so there is little agricultural activity. Some, however, is carried on in forest clearings usually on the southern margins of the forests. Hardy varieties of cereals (oats, barley and rye), potatoes and quick-growing types of green vegetables are cultivated.

Many of these forest regions are rich in supplies of water-power and the generation of hydro-electric power has allowed electro-chemical and electro-metallurgical industries to be set up, e.g. Kitimat, in British Columbia (aluminium), and Rjukan, in Norway (nitrate fertilisers).

41. The influence of environment on life. Because of the cold winters human activities are largely seasonal in character and are mostly confined to the summer half of the year. For example, many people find seasonal work in the lumber camps. In the old days many trappers lived lone lives in the forest. Except where there are mining centres, population concentrations are few and the coniferous forest regions as a whole are sparsely peopled. Houses, as one might expect, are built of logs. The long, dark, cold winters are very trying and human health tends to suffer on this account. The Soviet Union, which is trying to colonise parts of these territories, is experimenting with towns, houses, transport and types of work in an attempt to solve the difficulties of living in these regions.

TUNDRA, POLAR AND ALPINE REGIONS

42. Location and extent. The tundra and ice-cap regions occur in high latitudes, roughly astride or poleward of the Arctic and Antarctic circles. Tundra lands in the northern hemisphere are found on the continental fringes of the Arctic Ocean. In the southern hemisphere there are few areas of tundra; Tierra del Fuego and South Georgia are the largest.

Many high mountain (Alps, Himalayas, Rockies, Andes) and high plateau (Tibet, Bolivia) areas have, because of their altitude, alpine conditions which approximate closely in certain respects to the tundra and ice-cap lands.

43. Climate. The tundra lands are characterised by the absence of any warm season and a long winter night. Temperatures are below freezing in winter, while in the warmest month they seldom average more than 10° C. Occasional daytime temperatures may be, however, quite high. The total precipitation is commonly less than 250 mm, coming mostly in summer. In winter the land is under snow. High winds may be experienced in winter, especially in coastal localities. The ice-cap lands are characterised by permanent frost.

In a broad sense the rise in altitude is equal to the increase in latitude, so that the tundra and ice-cap climates have their counterparts in the alpine pastures and the regions above the permanent snow-line. There are two important departures, however, which demonstrate the inadequacy of this analogy:

(*a*) While the intensity of insolation becomes less towards the poles, it increases at higher altitudes.

(*b*) Increased height brings with it rarefaction of the atmosphere.

44. Vegetation and animal life. There is, of course, no plant life in the ice-cap regions and very little animal life (other than marine life).

In the tundra lands in winter there is no plant growth since the ground is frozen and there is physiological drought. In late May the thaw starts but only the top few centimetres of the soil thaw. The lengthening days and increasing temperatures result in a rapid growth of vegetation and flowering perennials burst into blossom bringing colour to the landscape. The characteristic plants are mosses and lichens and creeping berry-bearing plants such as cranberry and crowberry. Trees are few except for such species as dwarf willows and birches. Numerous migratory birds use the tundra as breeding-grounds in summer while insects abound. Caribou, reindeer and musk ox live off the mosses and lichens. Fish and other semi-aquatic creatures, such as seals and walruses, are usually abundant, although in some areas they have been hunted almost to the point of extinction.

Mountains, if they are high enough, have a similar sort of vegetative growth to the tundra lands. In winter they are snow-covered but in summer the snow-line retreats and the mountain slopes are covered with grass and many small flowering plants.

45. Environmental handicaps. There are several handicaps to life in the lands of the polar margins:

(*a*) The harsh climate is not attractive to man and precludes agriculture.

(*b*) Permafrost (permanently frozen sub-soil) characterises these regions.

(*c*) The numerous insects are a nuisance to men and animals alike.

(*d*) Natural resources are very limited in kind, indeed, almost restricted to animals.

(*e*) Most of these lands are remote, isolated and rather in-accessible.

Mountain environments have climatic handicaps and communication difficulties, and soils are thin, stony and immature.

46. The influence of environment on occupations. The tundra is undeveloped commercially. There are one or two mining settlements but, apart from these, the tundra lands remain almost untouched. The native inhabitants tend to live either on the marine northern margins or on the southern forest margins. The former include most of the Eskimos who are skilled fishers and hunters of sea-mammals such as the seal; among the latter are the Lapps of northern Europe and other reindeer herders of Asia, e.g. the Samoyeds, Ostyaks.

On the high plateaux of the world a little cultivation of crops becomes possible in sunny, sheltered hollows and valleys but such crops are of the hardy type such as potatoes and quinoa (a cereal) grown on the inter-montane plateaux of the Andes. Here, too, as also on the Tibetan plateau, a scattered herbage provides food for pastoral animals such as llamas and sheep. The Alpine meadows of Europe provide summer pasture for cattle which are moved up to the *mayen* (May pastures) and *alps* when the snows have melted.

47. The influence of environment on life. The inhabitants of the tundra and ice-cap lands are few in number and mainly nomadic because the means of livelihood are limited. Whether the scattered tribes live by hunting and fishing or by primitive herding, a nomadic way of life is necessary, since the fishing and hunting grounds and the pastures quickly become exhausted; man must follow food supplies and seek fresh grazing grounds for his animals.

The Eskimos provide the supreme example of the close adjustment which may exist between man and his natural environment. Nowhere else in the world has man made more complete or more ingenious use of what a niggardly nature offers. The Eskimos' dwellings, clothes, implements and boats exhibit a wonderful adaptation to the environmental conditions. The quest for food and protection against the severity of the winter may be said to dominate the lives of the Eskimos.

In the case of the primitive herders, such as the Lapps, life may be said to revolve around the reindeer which provide most of his needs, e.g. much of his food, clothing, dwellings, tools and implements. Hair, hides, antlers and bones provide much of his raw material for clothes, tents, footwear, bedding, utensils and implements.

In Alpine regions man, again, is seasonably nomadic, for he moves up to the alpine pastures along with his stock, a movement known as transhumance.

PROGRESS TEST 11

1. Describe the natural vegetation and animal life of the equatorial rain forest. **(4)**

2. Explain the meaning of the following terms: selva, campos, taiga, tundra. **(4, 8, 38, 43)**

3. Describe the chief environmental handicaps to the development of the equitorial regions. **(5)**

4. Show how geographical conditions have influenced human life and development in regions of equatorial forest.   (6, 7)

5. Locate the chief areas of tropical grassland and briefly describe the climatic conditions of these grassland regions.   (8, 9)

6. Certain climates have distinctive alternations of wet and dry seasons. Which are these climates? Compare the summer and winter temperature conditions and the amounts of rainfall.   (9, 21)

7. The civilisation of the monsoon lands of Asia has been described as a "vegetable civilisation". What is meant by this? Do you think it an apt description?   (13)

8. Describe the natural vegetation of arid and semi-arid areas and show how the plants have adapted themselves to these dry conditions.   (16)

9. Describe the life of the nomadic herder.   (19)

10. Describe the natural vegetation associated with maquis and steppe.   (22, 27)

11. Show how the environmental conditions have influenced (a) the life and (b) the occupations of people living in lands having a Mediterranean type of climate.   (23, 24)

12. Trace the stages in the economic exploitation and development of the temperate grasslands.   (29)

13. What drawbacks do the coniferous forest lands have from the point of view of human occupation and development?   (39)

14. Explain the meaning of the following terms: permafrost, podsols, transhumance.   (39, 45, 47)

15. "The life of the Lapps revolves around the reindeer." Explain this statement.   (47)

APPENDIX I

Bibliography

Alnwick, H., *A Geography of Europe*, Harrap, 1948.

Beaver, S. T., "Ships and Shipping: the Geographical Consequences of Technological Progress", *Geography*, 1967.

Broek, J. O. M. and Webb, J. W., *A Geography of Mankind*, McGraw-Hill, 1973.

Brunhes, J., *Human Geography*, French edition translated by F. E. Rowe, Harrap, 1952.

Childe, V. G., *Man Makes Himself*, Fontana, 1966.

Childe, V. G., *What Happened in History*, Penguin, 1969.

Cosgrave, I. and Jackson, R., *Geography of Recreation and Leisure*, Hutchinson, 1972.

Dobby, *Monsoon Asia*, University of London Press, 1961.

East, W. G., *A Geography of Europe*, ed. G. W. Hoffman, Methuen, 1953.

Fleure, H. J., "The Geographical Distribution of the Major Religions", Societe Royale de Geographie d'Egypte, T.XXIV.

Gauld, W. A., *Man, Nature and Time*, Bell, 1946.

Houston, J. M., *A Social Geography of Europe*, Duckworth, 1970.

Janata, G., "Tourism", Ealing College of Technology, 1971.

Jones, E., *Human Geography*, Chatto & Windus, 1964.

Mackinder, H. J., *Democratic Ideals and Reality*, Penguin Books, 1944.

Money, D. C., *An Introduction to Human Geography*, University Tutorial Press, 1975.

Monkhouse, F. J., *Dictionary of Geography*, Edward Arnold, 1970.

Perpillou, A. V., *Human Geography*, Longman, 1977.

Philbrick, A. K., *This Human World*, Wiley, 1963.

Pounds, N. J. G., *Political Geography*, McGraw-Hill, 1972.

Randall, H. J., *The Creative Centuries*, Longman, 1944.

Robinson, H., *A Geography of Tourism*, Macdonald and Evans, 1976.

Robinson, H., *Monsoon Asia*, Macdonald and Evans, 1978.

Russell, R. J. and Kniffen, F. B., *Culture Worlds*, Macmillan, 1951.

Sauer, C. O., *Agricultural Origins and Dispersals*, American Geographical Society, 1952.

Stamp, L. D. and Gilmore, S. C., ed. *Chisholm's Handbook of Commercial Geography*, Longman, 1954.

Valkenburg, S. van, *Elements of Political Geography*, Prentice-Hall, 1940.

Examination Technique

1. Three essentials. The examination candidate, if he or she is to be successful, must:

  (*a*) obey the rubric, i.e. the instructions;
  (*b*) understand the questions asked; and
  (*c*) arrange his or her material satisfactorily.

2. Read the instructions very carefully. It is surprising how often candidates disobey the rubric. Frequently papers are divided into sections and candidates are asked to answer one or more questions from particular sections; make sure you answer the correct number of questions required in each section. If the paper asks the candidate to answer four questions, this means four and not five; you will receive no credit for questions answered in excess of the number asked. If the paper has a compulsory question, this must be attempted; it may, indeed, carry more marks than the remaining questions you are asked to answer. Again, if a question specifically asks for sketch-maps or diagrams these should be given, for it is very likely that a proportion of the marks for that question has been set aside for the maps/diagrams. Try to make your sketches neat, reasonably large, accurate and to the point and do not waste valuable time merely embellishing them. Do not repeat by means of a map what you have already said in words. A map should be drawn to save you writing time or to *add* to your verbal description.

3. Read the questions carefully. *Make sure you understand what the examiner is asking.* Study the question carefully; it is helpful to underline the key or salient points in the question. Note the significance of *describe, discuss, explain, account for, compare* and *contrast:* each has a different meaning. For example, there is very little point in giving a description of the distribution of population in Australia if the question asks for the causes underlying the population distribution pattern. If the question asks for "either/or" this means one or the other, not both; if a question asks for "two of the following" it means two, not three. These points seem so obvious and yet it is surprising how many students err on small points of this kind. Such stupid mistakes can mean only one thing—the candidate has not read the question carefully. Again, avoid superfluous and irrelevant "padding"; by padding, the student is not fooling the examiner, only himself. Obey the injunction: "Answer the question, the whole question and nothing but the question."

4. Arrange your material carefully. Organise the material of your answer in an orderly, systematic and logical way. It is a good plan, after having read and understood the question, to jot down your ideas on a spare piece of paper or in your script and to arrange the main headings under which you intend to answer the question; see that you cover all the points. Be precise; give figures, if possible; give appropriate examples. Avoid such meaningless phrases as "the right type of soil", "a good climate", "in many other areas", etc. Locate areas precisely. Avoid making generalised statements which are untrue, e.g. "Monsoon Asia is a thickly populated region", "The countries of Africa are inhabited by Negroes". Sometimes a question is set which involves comparison and contrast: in such cases, take a point at a time and consider the similarity or dissimilarity. Two quite separate and distinct descriptions do not necessarily constitute a comparison or contrast. Moreover, if the question specifically asks for a comparison, the candidate may be penalised if he does not fulfil the instruction.

5. Presentation. Finally, and this should not need emphasising, write legibly and neatly and use good English, paying some respect to grammar, punctuation and spelling. Avoid using slang, abbreviations and ungeographical expressions. Exemplify and amplify your statements. The quality, not the quantity, of your answers matters to the examiner. Allot your time carefully so that you do not over write on any particular question. The good candidate and the one who scores heavily in examinations is the one who can give that little bit extra, whether it be in information, argument or sketch-maps, which places him above the general run of candidates.

Examination Questions

1. For any *one* continent you have studied draw a sketch-map to show the distribution of population. Explain the factors which have influenced this distribution. (*Welsh Joint Education Committee*)

2. For any large country in the tropics which you have sudied explain: (*a*) why it is backward; and (*b*) what measures are being taken to remedy this. Illustrate with a sketch-map. (*Institute of Bankers*)

3. "Nearly half the land surface of the earth cannot support human life." (*a*) In what parts of the world are these areas situated? (*b*) What geographical conditions render them useless to man for food production? (*London*)

4. Towns often spring up at: (*a*) the limit of ocean navigation; (*b*) at bends in a river; and (*c*) at the confluence of two rivers. Name and show, with the aid of diagrams, *a town of each type* and describe its economic growth. (*Northern Counties*)

5. The growth of "million cities" is largely a modern phenomenon. Can you explain the growth of such towns of immense size?

6. Name *two* distinctive areas within the tropics, one with a low and the other with a high population density. Describe the location of each and try to account for its population density. (*London*)

7. Agriculture in some countries is underdeveloped. (*a*) For each of *two* named areas, explain why this is so. (*b*) In what ways are the more advanced countries helping underdeveloped countries to improve their farming? (*Associated Examining Board*)

8. What is meant by the term "ethnic group"? What criteria may be used to differentiate the chief so-called "racial groups" of the earth?

9. Explain the meaning of the following terms and give examples of each: nucleated village; linear village; wet-point village; dry-point village.

10. "No amount of external aid alone will solve the problem of the underdeveloped countries; only self-help will bring the solution." Discuss.

11. Suggest new sources and methods of food production which will help solve the problem of world hunger.

12. "Culture and economic progress have never been achieved where environment is easy." Discuss. (*Oxford & Cambridge*)

13. "Cultural features may influence man's activities and outlook." Justify this statement by quoting some examples.

14. Draw sketch-maps and cite a specific example of the following types of towns: (*a*) a gap town; (*b*) a bridge town; (*c*) a town occupying a defensive site; and (*d*) a river confluence town. (*Northern Counties Technical Examinations Council*)

15. From your reading about (*a*) the pygmies of the Congo, and (*b*) the Greenland Eskimo, show how they have adapted their lives to the environments in which they live.

16. Consider, with reference to any part of Great Britain, the geographical factors that have influenced the distribution, size, and plan of rural settlements. (*Oxford & Cambridge*)

17. Which different building materials are used in the town or village where you live? Where do they originate and for what particular purposes are they used?

18. Which geographical factors do you consider relevant to the study of the problem of closer political union in Europe?

19. Attempt to justify the claim that *either* Latin America *or* Monsoon Asia forms a distinct culture realm.

20. Write geographical notes on *two* of the following subjects: (*a*) spring line settlements; (*b*) ports which have declined in importance; (*c*) gap towns.

21. Name an example, and explain briefly what is meant by *three* of the following: confluence town, dormitory town, gap town, holiday resort, naval base. In each case, draw a sketch-map to illustrate the particular nature of the town you have named. (*London*)

22. To what extent do you think disease, malnutrition and illiteracy are related?

23. Write explanatory notes on three of the following: *Geopolitik*, soil erosion, kwashiorkor, tsetse fly, dormitory towns, artificial capitals.

24. Describe the life of a typical farmer in *two* of the following areas and show how his occupation is influenced by the natural conditions of his environment: the Saharan oases, the Swiss Alps,

the Argentinian Pampas, southern Japan, North Island, New Zealand. (*Associated Examining Board*)

25. Compare the factors which have influenced the development of livestock farming and intensive crop cultivation in selected areas of the developed world. (*Associated Examining Board*)

26. What is meant by the terms "capital city" and "federal capital" and name an example of each. (*b*) With the aid of a sketch-map, write an account of the functions and urban patterns of a capital city you have studied. (*c*) Explain why either capital cities or federal capitals are not always the largest cities in their states. (*Associated Examining Board*)

27. Write explanatory notes of, and give examples of, *three* of the following: urban agglomeration, urban sphere of influence, hypermarkets, megalopolis.

28. (*a*) Name and delimit an area which you have studied in the field. (*b*) With the aid of sketch-maps and diagrams discuss the results of your study of *one* of the following: rural land use and farming practices; industrial activities; communications networks; shopping facilities. (*Associated Examining Board*)

29. (*a*) Distinguish between "recreation" and "tourism". (*b*) Select one important tourist destination, either in Britain or abroad, and analyse the chief factors contributing to its success as a tourist destination.

30. Discuss the changing fortunes and importance of the Middle Eastern area during the twentieth century.

31. From the less developed world, select one country you have studied and describe the geographical factors which appear to cause the population density in certain areas to be (*i*) dense, (*ii*) moderate, and (*iii*) sparse. (*Associated Examining Board*)

32. Describe, and attempt to assess, the value of the oceans and the ocean beds as a resource for the future.

33. Write an essay on international trade describing especially (*i*) the principal trading flows, and (*ii*) the growth of trading blocs.

34. Explain the various reasons for the establishment of new towns in the United Kingdom and quote examples of the various types of new towns which have emerged since the Second World War.

35. What is meant by the term "natural resources"? Outline, and contrast, the natural resources of any two countries of your choice.

36. Many of the present-day world problems are caused by the inequality in standards of living as between various countries.

(*a*) What determines the standard of living in a country? (*b*) What efforts have been made over the past thirty years to assist the underdeveloped countries of the world?

37. Write an essay on the domestication and use of animals.

38. Write an essay on the Eskimos to illustrate (*a*) how their lives were adapted to the particular environmental conditions under which they lived, and (*b*) how their lives have changed during the past three generations.

Index